£24.00

EUROPEAN COORDINATION CENTRE

FOR RESEARCH AND DOCUMENTATION IN SOCIAL SCIENCES

AUTOMATION AND INDUSTRIAL WORKERS

Other Publications of the Vienna Centre

BAEYENS, H., DREWETT, R. & VUKOVICH, G.
Urban Change and Societal Developments

CAO-PINNA, V. & SHATALIN, S.
Consumption Patterns in Eastern and Western Europe

FORSLIN, J., SARAPATA, A. & WHITEHILL, A.
Automation and Industrial Workers, volume 1, part 2

PENOUIL, M. & PETRELLA, R.
The Location of Growing Industries in Europe

SZALAI, A. & PETRELLA, A.
Cross-National Comparative Survey Research: Theory and Practice

VAN DEN BERG, L., DREWETT, R., KLAASEN, L., ROSSI, A. & VIJVERBERG, K.
Elements of a Theory of Urban Development and Policy

AUTOMATION AND INDUSTRIAL WORKERS

A Fifteen Nation Study

VOLUME 1

Part 1

Edited by

JAN FORSLIN, ADAM SARAPATA
and
ARTHUR M. WHITEHILL

In collaboration with

FRANK ADLER and STEPHEN MILLS

PERGAMON PRESS

OXFORD · NEW YORK · TORONTO · SYDNEY · PARIS · FRANKFURT

U.K.	Pergamon Press Ltd., Headington Hill Hall, Oxford OX3 0BW, England
U.S.A.	Pergamon Press Inc., Maxwell House, Fairview Park, Elmsford, New York 10523, U.S.A.
CANADA	Pergamon of Canada, Suite 104, 150 Consumers Road, Willowdale, Ontario M2J 1P9, Canada
AUSTRALIA	Pergamon Press (Aust.) Pty. Ltd., P.O. Box 544, Potts Point, N.S.W. 2011, Australia
FRANCE	Pergamon Press SARL, 24 rue des Ecoles, 75240 Paris, Cedex 05, France
FEDERAL REPUBLIC OF GERMANY	Pergamon Press GmbH, 6242 Kronberg-Taunus, Pferdstrasse 1, Federal Republic of Germany

First edition 1979

British Library Cataloguing in Publication Data

Automation and industrial workers.
Vol.1. Part 1
1. Automation - Economic aspects - Europe
2. Automation - Social aspects - Europe
3. Labor supply - Europe
I. Forslin, Jan II. Sarapata, Adam
III. Whitehill, Arthur Murray
IV. European Coordination Centre for Research and Documentation in Social Sciences
331 HD6331.2.E9 79-40811
ISBN 0-08-023339-2

In order to make this volume available as economically and as rapidly as possible the authors' typescripts have been reproduced in their original forms. This method has its typographical limitations but it is hoped that they in no way distract the reader.

Printed and bound in Great Britain by William Clowes (Beccles) Limited, Beccles and London

Contents

List of Participants

1. National teams

AUSTRIA

Participants	Institutions
Co-chairman:	
Mr. Erich Hofstetter, MP, General Secretary	Austrian Federation of Trade Unions
Dr. Alfred Wakolbinger, General Secretary	Austrian Chamber of Commerce
Members:	
Ing. Hans Fuchs	Federal Ministry for Science and Research
Dr. Sigurd Hoellinger	ditto
Mr. Kurt Prokop, Secretary	Austrian Federation of Trade Unions
Dr. Georg Wieser, Consultant	Austrian Chamber of Commerce; Institute of Sociology, University of Vienna
Dr. Gottfried Winkler	Scientific Dept., Austrian Chamber of Commerce
Director	
Dipl. Ing. Fred Margulies, Secretary	Union of Salaried Employees, Austrian Federation of Trade Unions

Fieldwork assistants:

Dr. Irene Geldner	Austrian Chamber of Labour
Ing. Walter Hackl-Gruber	University of Technology, Vienna
Mag. Paul Kolm, psychologist	
Ing. Rudolf Ruprecht, engineer	Optical Industry, Vienna

CZECHOSLOVAKIA

Participants	Institutions
Prof. Radovan Richta	Czechoslovak Academy of Sciences, Institute for Philosophy and Sociology
Dr. František Charvat	ditto
Dr. Jan Řehak	ditto
Mr. Milan Tucek	ditto
Mr. Yaroslav Hudecek	ditto

DENMARK

Participants	Institutions
Researcher and coordinator	
Mr. Flemming Agersnap,	Institute of Organization and Industrial Sociology, Copenhagen School of Economics and Social Science
Research fellow	
Mr. Poul H. Pedersen	ditto
Research assistants	
Mr. Steen Hvid (1973-76)	ditto
Mr. Mogens Larsen (1973-76)	ditto
Mr. Peter Nielsen (1973-76)	ditto

FINLAND

Participants	Institutions
Researchers	
Dr. Liisa Rantalaiho docent in sociology	Finnish Academy, Social Science Council, University

	of Tampere, Institute of Sociology
Lic. Soc. Sc. Tuula Heiskanen	ditto
Fieldwork assistance:	
M. Soc. Sc. Thelma Vuonokari	University of Tampere, Institute of Sociology
M. Sc. (Admin.) Tapani Korhonen	ditto
M. Soc. Sc. Rauno Parikka	University of Helsinki, Institute of Social Psychology
M. Soc. Sc. Markku Kuokkanen	University of Tampere, Institute of Sociology
	Organisational support:
	Finnish Metalworkers' Union
	Finnish Metal Trade Employers' Association

FRANCE

Participants	Institutions
Principal Researchers:	
Prof. William Grossin	University of Nancy
Dr. Yvette Lucas	National Scientific Research Centre (CNRS) University of Toulouse, Sociological Research Centre
Researchers	
Mrs. Chantal Benayoun	ditto
Mrs. Monique Membrado	ditto
Mrs. Emmanuelle Soula	ditto
Mr. Gilbert Faydit de Terssac	ditto

FEDERAL REPUBLIC OF GERMANY

Participants	Institutions
Dr. Guenter Friedrichs	Metalworkers' Union of the Federal Republic of Germany, Automation Department

Mr. Raimar Birkwald	Ditto, Collective Bargaining Department
Dr. Hans Otto Messedat	Confederation of German Employers' Association
Dr. Dieter Mertens	Institute of Employment Research of the Federal Employment Institute (Institut fuer Arbeitsmarkt und Berufsforschung - IAB - der Bundesanstalt fuer Arbeit)
Dipl. Ing. Erhard Ulrich	ditto
Ing. (grad.) Gerhard Dörfer	ditto
Ing. (grad.) Manfred Lahner	ditto
Dr. Hermann Saterdag	Institute for Market and Social Research (INFRATEST)

GERMAN DEMOCRATIC REPUBLIC

Participants	Institutions
Dr. Rosemarie Winzer	National Committee for Sociological Research of the GDR
Dr. habil. Manfred Lötsch	ditto
Dr. Frank Adler	ditto/European Coordination Centre for Research and Documentation in Social Sciences
Dipl. oec. Klaus Fischer	ditto
Dipl. Psych. Rainer Schubert	ditto
Dipl. Phil. Dietmar Wittich	ditto

GREAT BRITAIN

Participants	Institutions
Mrs. Enid Mumford	University of Manchester, Manchester Business School
Mr. Stephen C. Mills	ditto/European Coordination Centre for Research and Documentation in Social Sciences
Mr. Paul Lloyd	The University of Aston, Management Centre.

HUNGARY

Participants	Institutions
Principal Researchers:	
Dr. Lajos Héthy, C. sc.	Research Institute of Sociology, Hungarian Academy of Sciences
Dr. Csaba Makó, C. sc.	ditto
Fieldwork assistants:	
Mr. László Boros, sociologist	ditto
Mr. László Bélley, sociologist	ditto
Mr. János Bodó, psychologist	ditto
Mr. Csöppke Galambos, economist	ditto
Mr. Attila Gergely, sociologist	ditto
Mr. Tamás Gudlin, Dipl. Ing.	ditto
Mr. József Hegedüs, Dipl. Ing.	ditto
Mr. Sarolta Orbán, psychologist	ditto
Mrs. Eva Palkovics, sociologist	ditto
Mr. Zsolt Radics, Dipl. Ing.	ditto
Mr. Miklós Sághy, Dipl. Ing.	ditto
Mr. József Szabó, Dipl. Ing.	ditto
Other participants:	
Mr. György Horváth, Dipl. Ing.	Scientific Research Institute for Labour Safety, National Council of the Hungarian Trade Unions
Mr. Zsigmond Matolcsy, Dipl. Ing.	ditto
Mr. Gyula Nagy, Dipl. Ing.	ditto

ITALY

Participants	Institutions
Prof. Gustavo Iacono	University of Naples, Institute of Psychology
Dr. Guglielmo Bellelli	ditto

Mrs. Giovanna Petrillo	ditto
Mrs. Carmencita Serino	ditto

POLAND

Participants	Institutions
Prof. Adam Sarapata	Institute for Organization of Machine Industry

SWEDEN

Participants	Institutions
Researchers:	
Dr. Jan Forslin	Swedish Council for Personnel Administration
Mr. Jens Söderlund, psychologist AB Volvo	
Assistants:	
Mrs. Viveka Ferlander, B. A.	
Mrs. Ann-Marie Nobel, B. A.	
Mr. Ulf Svensson, B. A.	
Mrs. Barbro Zackrisson, B. A.	

U.S.A.

Participants	Institutions
Project director	
Mrs. Betty M. Jacob	Research Corporation, University of Hawaii
Principal Investigators:	
Prof. George W. England	University of Minnesota, Industrial Relations Center
Prof. Philip E. Jacob	University of Hawaii, Political Science Department
Prof. Arthur M. Whitehill	ditto College of Business Administration

Research Associate:	
Dr. Richard C. Pratt	University of Hawaii, Political Science Department
Research Assistants:	
Dr. Chung-Si Ahn (1975-77)	ditto
Mr. Garry Williams (1977-78)	ditto
Mr. Alex Farkash (1975-76)	University of Minnesota, Industrial Relations Center
Mr. Itzhak Harpaz (1976-77)	ditto
Management representatives:	
Mr. Edward N. Cole, President	General Motors Corporation 1970-74.
Dr. D. L. Landen, Director	Employee Research Division, General Motors Corporation
Union representatives:	
Mr. Irving Bluestone, Vice-President	United Automobile Workers (UAW)
Mr. Victor G. Reuther, Director	International Relations Dept. UAW (1970-71)
Mr. Nat Weinberg, Director	Special Projects and Economic Analysis, UAW (1972-73)
Mr. Herman Rebhan, Director (1973-75)	International Affairs Dept. UAW
Mr. Don Ephlin	Assistant to the President, UAW (1975--)
U.S.S.R.	
Participants	Institutions
Prof. Vladislav Ussenin	Academy of Sciences of the U.S.S.R., Institute of International Labour Movement
Prof. Boris Koval	ditto
Dr. M. N. Nochevnik	ditto
Prof. V. V. Krevnevich	ditto
Dr. V. V. Maslov	ditto
Dr. D. P. Podkosov	ditto

Dr. L.V. Shoustova	ditto
Dr. V. V. Kolbanovsky	Academy of Sciences of the U.S.S.R., Institute of Sociological Studies
Dr. G. N. Denisovsky	ditto
Dr. E. G. Sarovsky	ditto

YUGOSLAVIA

Participants	Institutions
Dr. Bogdan Kavčič (1972-75)	Self-management Research Centre of RU, Association of Slovenia
Dipl. psych. Valentin Jež	ditto
Dipl. soc. Ilija Globačnik	ditto
Dipl. soc. Frančec Kavčič	TU Association of Slovenia
Dipl. oec. Nada Mikić-Bulc	Self-management Research Centre of TU, Association of Slovenia
Mr. Rudi Peperko	Social attorney of self-management

2. Steering Committee

Prof. Flemming Agersnap (Denmark)

Dr. Ralph H. Bergmann (International Labour Organisation)

Dr. Jan Forslin (Sweden)

Dr. Gunter Friedrichs (Federal Republic of Germany)

Prof. William Grossin (France)

Dr. Lajos Héthy (Hungary)

Prof. Gustavo Iacono (Italy)

Prof. Philip E. Jacob (USA)

Dipl. psych. Valentin Jež (Yugoslavia)

Dipl. Ing. Fred Margulies (Austria)

Dr. Dieter Mertens (Federal Republic of Germany)

Mrs. Enid Mumford (Great Britain)

Mr. Kurt Prokop (Austria, Trade Union Coordinator - West)

Dr. Liisa Rantalaiho (Finland)

Prof. Radovan Richta (Czechoslovakia)

Prof. Adam Sarapata (Poland)

Prof. Vladislav Ussenin (U.S.S.R.)

Dr. Vladimir Veretenikov (U.S.S.R., Trade Union Coordinator - East)

Dr. Rosemarie Winzer (German Democratic Republic)

ex officio members of the Steering Committee

Dr. Frank Adler (Scientific Secretary of the project, Vienna Centre)

Mr. Stephen C. Mills (Director of the Vienna Centre)

Prof. Adam Schaff (President of the Board of Directors of the Vienna Centre)

3. Executive Committee

Dr. František Charvat (Czechoslovakia)

Dr. Gunter Friedrichs (Federal Republic of Germany)

Prof. Philip E. Jacob (USA)

Mr. Kurt Prokop (Austria)

Dr. Liisa Rantalaiho (Finland)

Prof. Vladislav Ussenin (U.S.S.R.)

ex officio members of the Executive Committee

Dr. Frank Adler (Scientific Secretary of the project, Vienna Centre)

Dr. Ralph H. Bergmann (International Labour Organisation)

Mr. Stephen C. Mills (Director of the Vienna Centre)

Professor Adam Schaff (President of the Board of Directors of the Vienna Centre)

4. Technical Advisory Group on Analysis:

Dr. Jan Forslin (Sweden)

Lic. Soc. Sc. Tuula Heiskanen (Finland)

Dr. Bogdan Kavčič (Yugoslavia) (1974-75)

Mr. Stephen C. Mills (Director of the Vienna Centre)

Dr. Richard C. Pratt (USA)

Dr. Jan Řehak (Czechoslovakia)

Mr. Marek Prussak (Poland)

Prof. Adam Sarapata (Poland)

5. Scientific Secretaries of the project

1. Dr. Risto Kolari, Finland, 1970-71
2. Mrs. Marianne Layer, Federal Republic of Germany, 1971-73
3. Mr. Stephen C. Mills, Great Britain, 1973-75
4. Dr. Frank Adler, German Democratic Republic, 1975-78.
   Dipl. Ing. Fred Margulies (Austria) acted as consultant to the project 1971-73.

Preface

This volume is the first in a series which will report on the results of the multinational study "Automation and Industrial Workers" co-ordinated by the European Co-ordination Centre for Research and Documentation in Social Sciences (Vienna Centre). Using co-operatively developed instruments, scientists from fifteen countries have probed the impact of advanced technology upon attitudes of workers, work content, and working conditions.[1] This broad representation illustrates that automation and its social implications are one of the most important theoretical and practical issues of today and the future.

In this initial volume the preliminary findings and analyses in four of the participating countries are presented. These contributions have not been narrowly structured nor have the authors followed a standard format. Rather, each contributor was encouraged to study the raw data for his country and, with a broad brush, paint the general picture which emerges. The conclusions reached are, of course, those of the writers, and neither represent a consensus among all project participants nor an official view of any given country.

The result is a collection of uniquely flavoured research papers. Because contributions to this volume were drawn from market-oriented economies and centrally and non-centrally planned economies, the influence of social, political, and economic system naturally becomes a variable to which considerable attention is devoted. Yet this is only one of the important determining factors examined by the authors in their comparison of automated and non-automated departments in major automotive enterprises in their respective countries.

This set of reports gives the reader an understanding of the phenomenon of automation within a number of national contexts. However, this volume is intended neither to provide a direct cross-national comparison nor to give an elaborate answer to the question of the influence of socio-economic system upon automation and its social consequences. The reader should be aware that since the data are not presented here in a strictly comparable form, cross-national comparison should be resisted at this stage. Furthermore, the reader is cautioned concerning terminology. Words are necessarily surrounded by a complex of associations, and since these differ between countries, therefore, it is essential that familiar words be read in the context of the specific national setting of each contribution.

The volume opens with an introductory part. Adam Schaff's introduction gives an insight into the "philosophy" of this multinational study. The author demonstrates the theoretical and practical relevance of the project's central problems-the impact of automation on the work and life of workers and the role of a country's socio-economic system in this relationship. Clarifying that different schools of thought give controversial answers to these key issues of our epoch, he underlines the project's realistic "modesty", being neither able nor willing to give definite answers. R. H. Bergmann (IILS) deals with the background, some specifics of the study, and its organisation. From his experience as chairman of the project's Steering Committee, he gives an insight into those necessary elements. This helps the reader understand the whole framework and development of the project.

Then a methodological paper is presented by V. Ussenin and M. Notchevnik of the Soviet Academy of Sciences, which describes several unique aspects of international comparative studies. The authors discuss the special nature of cross-national studies as a process and, in so doing, reveal both the necessity for and the opportunities and limitations of such research and its applications to the present project.

Following these introductory papers, the first of the contributions is the report from Sweden. It quickly rejects the primary importance of automated as opposed to non-automated technology in determining the attitudes and behaviour of industrial workers. Instead, the authors focus upon differences found between traditionally organised work units and one demonstrating an innovative approach to work organisation. We have here a critical assessment of the customary forms of social organisation in the work process of modern industry and an enthusiastic picture of the success of an experimental form in a Swedish plant.

The United States report is the second contribution from the market-oriented countries, and again offers a different approach to the common topic. In addition to reporting the responses of American workers to the most important single item and composite variables in the study, a number of special problems are addressed: the key role of job security in influencing acceptance of automation; the existence of alternative sources of "profiles" of worker motivation; some causes for the problem of alienation among youthful workers; and the role of a core of activists.

Adam Sarapata's paper presents the first material from a socialist country-Poland. Here, apart from analysing the impact of automation on worker's attitudes and aspirations, the author provides a deeper understanding of this relationship against the historical background and the work culture of the Polish nation. Emphasis is also put on methodological issues interesting for a wide public as well as the basic problem area of workers' attitudes towards technological development and the psychological reasons for the diversity of expectations in this connection.

Finally, the contribution from the Soviet Union closes the initial selection of reports. Considerable emphasis upon the beneficial impact of automation in relation to a wide variety of issues is expressed by the Soviet authors. This is a comprehensive analysis with much interesting data. Perhaps of special interest is their conclusion that the social and economic system is of overriding significance in determining the various outcomes of advanced technology.

So while each team's contribution to this volume deals with data drawn from common instruments dealing with a single topic, a remarkable variety of interpretation and selection in analysis has been demonstrated. Although seen from different angles, an almost unanimous conclusion in these reports is the rejection of the idea of technological determinism, i.e. that automation is a blind force both independent of its social content and unaffected by its actors.

The volume closes with a part containing the five research instruments, short descriptions of the optional and associated studies of the project, and a list of background papers. These allow the reader to become acquainted with the framework of the entire study and to understand the common sources of the information presented.

This collection is the first volume of book I and will be followed by an additional selection of reports by the participating teams in a second volume. This, in turn, will be followed by a comprehensive book of comparative findings as well as by the other books, monographs, and articles arising from this ambitious international research programme. It is the editors' hope that this first publication will both introduce the reader to the project and whet his appetite for further information as additional publications appear.

On behalf of the editorial group, special thanks are due to the Polish and Yugoslav sponsors for their generous hospitality and excellent facilities which made the editing of this volume possible.

Editors:

Jan Forslin-Sweden
Adam Sarapata-Poland
Arthur M. Whitehill-USA
Scientific Secretary of the Project:
Frank Adler - Vienna Centre (GDR)

Zabarow, Poland, July 1976
Dolejnske Toplice, Yugoslavia, September 1976

NOTES

1 Participating countries include Austria, Czechoslovakia, Denmark, the Federal Republic of Germany, Finland, France, the German Democratic Republic, Great Britain, Hungary, Italy, Poland, Sweden, the United States, the Soviet Union and Yugoslavia.

PART I

1. Introduction

ADAM SCHAFF

"Automation and Industrial Workers" is the topic of one of the most ambitious and important studies so far carried out by the European Co-ordination Centre for Research and Documentation in Social Sciences (the Vienna Centre). This information is a fitting and necessary initial perspective for the reader of the first volume of a series of publications containing the results of this research.

Research teams from fifteen countries have worked for 5 years to accomplish this project. But the magnitude and weight of the project is due not only to the number of participating countries involved, nor is it only because this single label actually covers a number of research projects[1]. What is most important is the complex character of these investigations in terms of their theoretical goals and of the instruments and research techniques used. Here the project makes a real contribution to the development of cross-national research. These aspects will be dealt with more fully in the two following articles by R. H. Bergmann and V. Ussenin, which introduce both this volume and, in a sense, the whole series.

However, as President of the Centre's Board of Directors, I should like to focus on the "philosophy" of this project. For it is this philosophy which is of primary and vital importance to the Vienna Centre as a sponsor and co-ordinator of this research programme. What we are talking about here are the theoretical and practical expectations of the Centre regarding the anticipated research results and the theoretical standpoint determining the choice of the principal hypotheses of this study.

The Vienna Centre was originally founded by UNESCO with the aim of forming a common ground for co-operation between social scientists from countries with differing economic and political systems. Thus, some years ago, the Institute of International Labour Movement, one of the institutes of the Soviet Academy of Sciences, inquired whether the Vienna Centre could organise a comparative research on the social consequences of automation in the highly industrialised countries of the world. After agreeing to undertake this, two main tasks had to be solved:

(1) to form an international research team (i.e. to assure the participation of research groups from the different countries); and then
(2) to elaborate the design of the research project itself.

The first task-painstaking and laborious as it was-is of less interest to us in the present context. Briefly, however, it took almost 2 years for us to create a core group of international teams. To do this it was necessary to dispel political preconceptions concerning the possibilities of undertaking common research in such a sensitive field of inquiry, where fears of betraying certain industrial secrets are particularly strongly marked. And it was also necessary to convince research workers in a number of countries that despite considerable research on this topic in their countries, such a new inquiry was worth undertaking because of the distinctive feature of our research programme: the comparison of empirical data from countries with differing socio-economic and political systems. This would be possible for the first time because of the participation of research teams from socialist countries both with centrally and non-centrally planned economies alongside those representing the countries with a market economy, and because of the collaborative forms of inquiry elaborated by the Centre throughout its existence.

It is the second aspect of the problem which is of particular interest to us here, however: how the conception of the research project was developed.

This conception was modified several times until it finally reached a tangible but more modest basis than that we had originally set ourselves. Such a scaling down of the research effort is quite understandable considering the character of international comparative empirical research in the domain of social sciences: "modesty" in this respect was necessitated by factors such as the need to make the research tasks fully operational, the possibility of developing research instruments acceptable for all the teams involved, and, lastly, the limitation of material resources (staffing and budget) available for the execution of the research.

Initially, the scope of the problems considered was very wide. It was intended to make a specific diagnosis of the age in which automation of production processes aims to reach full automation, thereby giving rise to a number of major economic, social, and political problems. The participating researchers quickly came to the conclusion that a study of such scope and magnitude could not be realised at the present stage of social science development. So it was necessary to focus only on some parts of the problem in question. While the research goals were made more "modest", at the same time the original conviction was retained that the efforts undertaken could yield results that would contribute towards solving some really important and considerably more universal problems. And this conviction is strengthened if we assume the periodic repetition of the study, thereby enabling us to subsequently introduce a dynamic component.

One can hardly expect this study to give complete answers to the big questions of the age, particularly those relating to the advancement of automation. The researchers did not set themselves tasks of such

large dimensions, for these could not be realised on the basis of such limited samples of the population investigated, through the restricted scope of the problems studied, and having ruled out of the investigative process the dynamic element of automation in time. Nevertheless, even if this study is only part of the original, more extensive "philosophy", failure to take this initial aim into account can substantially obscure the understanding of its intentions.
The international research team faced two big questions which were formulated as the major hypotheses:

(1) whether and how automation affects the life of the industrial worker; and
(2) whether and in what way the socio-economic system affects the functioning and development of automation in individual countries.

As can easily be seen, these two problems are closely related to the larger diagnostic conception of the research referred to above, i.e. they are its logical consequence being, at the same time, the small bricks on which this large construction can be erected. Hence, the obvious importance of the two problems. Since there exist different schools of thought about the two cases respectively - different schools giving contradictory answers to the questions raised - they open a large area for testing the differing standpoints against the data obtained from the study, thereby making the research itself more theoretically interesting.

These problems, and the hypotheses related to them, were put forward by the authors of the project who were fully aware of their theoretical and practical significance. Here a major influence was the range of theoretical concerns shown by many of the international teams, a number of whom were Marxists.

In relation to the first hypothesis, Marx had made a striking prognostication long before automation existed.

Marx, admittedly, treated hired labour (Lohnarbeit) as a form of alienation. The worker feels himself as a human being only after having finished his paid labour, when he is really reduced to animal functions satisfying the need of hunger, sleep, reproduction, etc. It is paradoxical in Marx's view that work, being a specifically human attribute and function, is felt by the worker as inhuman due to alienation while, on the contrary, the animal functions tend for him to take on human features. The solution, as Marx saw it, was - taken as given the consequences of a social revolution - in the reduction of working hours until gradually reaching the point of their full liquidation as a result of automation. Marx displayed this standpoint in his early Economic and Philosophical Manuscripts (1844), and repeated it in a modified form in the third and last volume of The Capital. There he made his claims more pronounced by noting additionally that the situation described refers to all societal forms.

Since the so-called creative work - necessary as it may be - is socially insufficient, for work which produces man's means of subsistence is indispensable, the only reasonable conclusion on the basis of Marx's attitude to hired labour as alienated work is - on the basis and as a result of social revolution - the substitution of human labour by machines. And the extreme point of this development would be full automation of production processes.

And so Marx consistently develops his ideas towards exactly this point. In the first outline of The Capital, the so-called Grundrisse (1857),[2] Marx explicitly anticipates full automation as the final point of the process of consecutive metamorphoses of the so-called fixed capital. In addition, Marx deals with the role of science as a productive force and with the altered function of the worker in the fully automated production process. Marx's argumentation on the subject, particularly that contained in pages 690-695 of the English translation[3], strikes us as the most perceptive prophecy of the development of automation and its social consequences.

In the Marxist view it is obvious that the problem of how automation affects the life of the individual worker not merely refers to psychological issues such as job satisfaction. It is a part of the wider and more basic issue of making human life more human, not only in the sense of "work humanisation" but in the sense of a possibility to substitute creativity for labour and of the social consequences of that process.

The answer to the second question, i.e. whether the economic and political system of a given nation affects the automation phenomena, stands out even more clearly. It is possible to identify two main schools of thought here. One that rejects the existence of such an influence contending that modern technology determines the growth of a homogeneous, industrial or post-industrial society, and another that holds the opposite view maintaining that the introduction and development of new technologies and in particular the social consequences of this process are almost universally homogeneous but differ according to the systemic conditions in the individual societies. Thus the co-operation of researchers of different orientations was an additional advantage of our project. This co-operative approach required special working conditions and organisational forms which the Vienna Centre has developed over the years.

The variable of socio-economic system played a central role in the total structure of our research. Again we had no illusions and did not intend to give a full or definite answer. Therefore, the data are regarded as the basis for different generalisations coming from different schools of thought. We made it our rule that every problem in the comparative analysis should be dealt with by at least two authors representing different schools of thought. And that in discussions published at the end of each problem, all the researchers concerned would be able to participate if they so wished.

Our pluralistic goals go even further in this respect. A series of publications consisting of monographs on particular countries has been designed. Another series will be that of comparative studies and there will also be a methodological series. In addition, authors not directly involved with the Centre's project are free to develop certain issues in a form of a monograph on the basis of our findings and to draw on the data collected in the project. This will enable fruitful discussions of the results obtained.

Finally, on behalf of the Vienna Centre's Board of Directors, I would wish to thank the workers, trade union officials, and management from all fifteen countries for making this study possible and providing the information on which it is based. Also, I should like to thank the scientific and sponsoring institutions, which enabled us to join together all our participating teams and provided working and meeting facilities at one time or another in all countries.

Despite the complexities which naturally arise during an investigation with fifteen participating countries from Europe and elsewhere, the atmosphere of co-operation was admirably good and the teams worked enthusiastically and very hard. I wish to convey my warm thanks to all the participating research teams and their leaders for their contribution to our common work. I should also like to acknowledge the help of the specialist groups that have dealt with particular problems of the research: e.g., the Technical Advisory Group on Analysis and the Trade Union Group.

Last, but not least, my thanks are due to the Centre's scientific and technical staff who co-ordinated the research with dedication and reliability.

NOTES

1 These are really four simultaneous studies. There are two Vienna Centre studies in the automotive and steel industries respectively; and there are two associated studies-one on computers and work design-the other on computerised management information systems. This volume deals only with national results of the first study.

2 This work ist still rather esoteric for the public at large due to its very belated publication - the first German version appeared in the USSR in 1939-1941 the the war annihilated the circulation, while the second publication appeared in the German Democratic Republic in 1953 but the English translation did not appear until 1973.

3 K. Marx, Foundations of the Critique of Political Economy, translated with a foreword by M. Nicolans, Penguin Books Ltd., Harmondsworth, Middlesex, England. (The pages referred to are to be found on pp. 582-590 in the German edition: K. Marx, Grundrisse der Kritik der Politischen Oekonomie, Berlin, 1953.)

2. The Project—History and Scope

RALPH H. BERGMANN

In agreeing to undertake an international research project on "Automation and Industrial Workers", research teams from countries with widely differing economic and social systems set out three broad viewpoints concerning the possible effects of modern technology on "man and society". They stated these in the "Final Research Design" (January 1973) as follows:

> "Some expect a general trend towards a new man with a positive creative attitude towards his work and with growing social activity. Others believe that for an increasing proportion of the working population automation will mean more monotonous work and greater job and social frustration. A third position is that automation will have virtually no effect on the people involved. It may change the nature of their work but not how they feel or behave in their social relations."

Addressing themselves to these alternative views, the research teams formulated a study to examine the nature and extent of the changes arising from automation in order to better understand the breadth of its "impact . . . upon the broader areas of social values and behaviour." In essence, the project provided for an empirical study of the effects of automation on working conditions, on work content and on worker attitudes in each of the participating countries. The findings were to be analysed according to a major hypothesis "that a country's political, economic, and social system will exert considerable influence both on the process of introduction and development of automation, and on its social consequences."

The work of focusing the research through an agreed research design required several meetings of participating national research teams over a period of almost 3 years. Subsequent efforts, not even yet completed, have flowed from this structure.

It is the purpose of this paper to describe the project and its development, to identify the project's special characteristics, to explain the research methodology, and to sketch out the administrative arrangements.

1. BACKGROUND

After agreeing to organise the project, the Vienna Centre, utilising its contacts among academics and social science institutes, invited researchers from industrialised countries to formulate the project design in operational terms and to develop the research programme. A group of researchers first came together in a preliminary meeting in August 1971; they agreed that the project was worth undertaking and proposed further steps in its organisation.

The first full-scale or plenary meeting was organised at Erlangen (FRG) in January 1972. While background papers had provided a preliminary description of the project's objectives, it was clear at the meeting that widely different interests of the participating researchers would have to be reconciled. For some country teams, the question of worker attitudes was of paramount importance, while for others, changes in work content and/or working conditions were primary. From this diversity there arose an agreement to organise a "standard study" to collect information on all three issues, with a concurrent agreement for the organisation of "optional studies". Such studies could be organised when there was sufficient interest, including the participation of researchers from countries with different economic and social systems ("East" and "West"), to examine any of the three subjects in greater detail, or to carry out the standard study in another setting.

It was also agreed from the very start that national research teams should seek to include employers and trade union officials in the research effort. At the first plenary meeting in Erlangen the trade union participants formed an international trade union group to advise on the project's formulation and to consider overall union interests in the research. This union group has assisted in the preparation of the research instruments, particularly that set of questions designed to elicit the views of union officials, and in other ways has provided guidance from the worker viewpoint. It is currently considering ways in which the research findings may be utilised for union policy making.

The formulation of the research design and the development of the research instruments were carried out in a series of meetings in 1972 and 1973. Following the full-scale meeting in early 1972, a representative working group met in Semmering (Austria) in mid-1972 to prepare a preliminary draft of various instruments. These were submitted to a second plenary session in Moscow in October, which formally authorised the undertaking of the project. In January 1973 a working group met at Neuwaldegg (Austria) to finalise the research design and research instruments. Pre-testing was carried out in the spring of that year.

In August 1973 a further working group was organised in Pezinok (Czechoslovakia) to evaluate the pre-test results. Almost all countries participated in the pre-test exercise, undertaking interviews with a total of 356 workers as well as with management officials and trade union leaders. The pre-test responses had been coded for computer handling and processed in Prague, and the print-outs were used to determine how well the various questions had worked in practice. Using various statistical tests for reliability, and taking account of conceptual and interpretative problems reported by national teams, the Pezinok group eliminated some questions and variables, while modifying others. The recommendations of this group were then passed to the third plenary session which met at "Wingspread" in Racine, Wisconsin (USA), in November of that year.

As the final step before the fieldwork, a group met in Poland in January 1974 to review each item in the research instruments. Their task was to finalise the instruments and formulate a manual of field research procedure which would enable each national research team to proceed in a standardised fashion. The bulk of the field research was carried out in the spring and summer of that year. Each national team sent their data on punch cards to the Computer Centre of the Institute for the Organisation of Machine Industry, in Warsaw, where, thanks to the generosity of the Institute, the project was able to set up its common working data bank.

First computer results were discussed at a working group meeting in mid-1975 in Lauf (FRG) and modifications and adjustments required further technical meetings in 1975 and 1976, including a special working group meeting in Tampere (Finland), which considered questions arising from the workplace observations. Most difficulties were settled by the April 1976 working group meeting in Vienna, and later that month each team received a tape (or print-out) providing, for each country in the project, a set of correlation matrices, contingency tables, and univariate statistics covering the study variables.

Another important outcome of the Vienna meeting was a decision concerning the first common publication of the project's results. This was to be in the form of a collection of articles based on each country's own data and highlighting the points of most interest to that country's team. The meeting further elaborated a general programme for cross-national analysis and publication.

2. SPECIAL FEATURES OF THE PROJECT

In the decade of the 1960s, "automation" was the subject of both extensive and intensive research. Most studies dealt with the employment effects of the new technology, including especially such issues as the possibility of unemployment, and the implications for skill requirements and systems of education. But, in addition, a number of studies considered the three topics which were examined in this study.

Some of the earlier studies were based on economy-wide or industry data and sought to describe broad consequences of the new technology; other studies were grounded in empirical, plant-level research. The goals, methods, and conclusions of those prior studies provided guides for the implementation of the current project, and the research structure benefitted also from previous investigations of worker attitudes and from publications providing systematic descriptions of work content and working conditions.

It is not possible to summarise here the many national and international automation studies which have been published[1] nor to indicate comparisons between the findings of this project and such reports. Each national team, however, has had the responsibility of considering previous studies in their country as they analysed their data and developed their findings.

A consideration of those aspects of this project which set it apart from other studies of new technology, shows that the differences are both procedural and substantive in character. On the procedural front it should be noted that the research was undertaken in fifteen countries according to a common methodology, using identical research instruments, during approximately the same period of time. It thus permits a structured comparison of the effects of a common technology in those countries.

Secondly, the project was carried out in countries with a "centrally-planned economy" and countries with a "free-enterprise economy", with the express purpose of determining whether the effects of "automation" differ according to a country's economic and social system.

Another feature of the research has been the tripartite organisation of the study. Management and trade union representatives have been involved with researchers in each step of the programme at both the national and international level. This has kept the study relevant to problems of practical as well as theoretical significance. In some countries it enabled the survey to be conducted in plants which were previously inaccessible. Trade union advisers, with their links to the workers being studied, have assisted in the interpretation of the findings, and union representatives from a number of countries are currently examining ways to co-ordinate their national approaches to issues raised by the new technology.

Furthermore, by bringing together research teams from six countries with a "planned"economy and from nine countries with market economies, the project provided an opportunity for "East-West" co-operation on a scientific endeavour. This was particularly noteworthy since the subject matter involved social issues and, therefore, touched on potentially sensitive questions. Most examples of other co-operative research between East and West involve natural and technical sciences.

Finally, it is noteworthy that the participants from all fifteen countries unanimously agreed that: "Each national team . . . shall have full access to the data and analyses of all . . . participants for purposes of cross-national comparison and analysis."

With respect to the substance of the research itself, it is again noteworthy that this study links together the observations made by outside experts at each workplace with the attitudes of the workers at just those workplaces, in contrast to many other studies which have examined worker attitudes towards automation and other job-related issues, or have been limited to observations of the impact of new technology on work and working conditions.

Another special feature lies in the study's focus not only upon work-related issues but also upon broader implications such as a worker's commitment to social goals and participation in civic affairs.

Furthermore, whereas in most studies the term "automation" refers to the global technological level of a plant, in this study the level of automation has been measured at the individual workplace and can thus be related to the attitudes of the specific worker.

3. PLAN OF RESEARCH

Although all participants agreed that the study was to examine "automation", the term had widely different meanings for different groups. In light of the basic decision that this would be an empirical study, it was necessary to select a particular type of modern technology whose effects could be observed and analysed. Furthermore, it was necessary that this "automation" be in sufficiently widespread use that it would be found in all participating countries, and that more than one establishment in a country could be studied. Within these constraints, the transfer line for the machining of automotive engine blocks was selected, and, with only few exceptions this technology was the kind of "automation" whose effects were examined.

Participants recognised that this solution was hardly perfect. They understood that in different countries (and plants) the time interval between the introduction of the technology and the conduct of the study would vary widely and thus possibly influence the findings. Diversities between transfer lines, and differences in the extent of the total work process performed by different transfer lines, were also likely to affect the results, and these points were noted for consideration when international comparisons were to be attempted.

It was also necessary to choose a second type of operation so that comparisons could be made between working conditions and worker attitudes in two different technological settings. For this purpose, the study focused on similar machining operations on multi-purpose manually controlled machine tools. The basic structure of the study called for a comparison between an automated unit (department or shop) and a non-automated unit. Data was collected from two, three, or four plants in order to control for plant-level differences.

The principal variable in the research design was automation, while the major dependent variables were (a) work content, (b) working conditions, and (c) worker attitudes. Important intervening variables such as characteristics of the worker, the plant, management system, and union were also included. All of this was viewed within the context of the country's economic and social system.

When the data were analysed it became apparent that work organisation was one more essential intervening variable which had to be taken into account. This decision was based on the observations of several research teams, which noted that differences in work organisation in plants being studied had an influence on worker attitudes.

Given the structure of the research design and the objectives of the study, it was clear that a number of separate instruments would be required for data collection. The following five instruments were developed and pre-tested over a period of 2 to 3 years, with the understanding that each would be used in the same way in all participating countries.[2]

I. Social and economic system of the country and its influence on technological changes and their social consequences.

II. Information obtained from management at the plant level.

III. Information obtained from trade unions and other worker representatives.

IV. Work process and workplace observations

V. Workers' questionnaire.

These instruments reflected the study's need to have data on several interrelated levels. Data from the last two sources (the workplace observations and worker interviews) were linked through the computer while preserving the anonymity of the respondents. In all of the interview instruments some common themes appeared, especially attitudes toward automation and its consequences for work and workers, and perceptions of participation in decision making at the plant level.

Instrument I was to provide a description of the country's social and economic system. It calls for statistical, historical, and descriptive documentary data, covering especially ownership, regulation, and management of industry, labour legislation, social security provisions, and education.

Instrument II was designed to secure two types of data from management sources. It contained questions about plant characteristics, such as its age, products, size of workforce, wage system, shift system, etc. Secondly, it sought the views of top management on various issues connected with automation.

Instrument III was prepared for interviews with trade union leaders and other worker representatives. This generally followed the line of questioning of the second part of Instrument II and picked up some additional points from the workers questionnaire (Instrument V). Furthermore, its questions were designed to be asked separately of national union leaders in the automotive or metal-trades industry and of plant-level officers. It thus had the potential of providing comparisons between the views of different levels of union leadership as well as comparisons with the views of managment and of workers.

Instrument IV called upon an independent observer to examine the automated and non-automated units and, through a system of scales and codes, to record details about the tasks performed at, and the characteristics of, each workplace. Where necessary, the examination was to be done in consultation with work-standards engineers and foremen. Three reports were prepared: a layout chart of each department with a description of equipment, operations, jobs, etc.; a task analysis of each workplace; and a report on job requirements, working conditions, and level of automation for each position studied.

Instrument V consisted of the series of questions to be asked of workers. During its preparation, the research groups made every effort to choose terminology which could be translated into other languages and to restrict the questions to concepts which had significance in the participating countries. Many hours were devoted to overcoming problems associated with the scaling of responses, the construction of a measure of the variable from replies to a group of questions, and the development of internal checks - all issues which were complicated by the multi-country dimension of the project. In some instances, questions appeared to have political implications, and special efforts at mutual understanding and goodwill were required to develop acceptable compromises.

The development of an agreed set of instruments required extensive discussion and a pre-test. Items were included if they provided information on a research variable and had relevance in a number of countries. Decisions were based on experience in previous national or cross-national studies of worker attitudes, and on an analysis of the pre-test results.

In its final form the workers questionnaire included questions covering several subject areas. The workers were asked how much their work provides variety, a chance to develop one's abilities, and the opportunity to use one's knowledge. They were asked how much scope their work gives to social relationships with co-workers, what were their relations with their supervisor, how much opportunity they had for upgrading and improving their skills and education, and their opinion about pay, job security, physical working conditions and their mental and physical work load. Paralleling these questions on the extent of various job characteristics were other questions to determine the importance of each characteristic for the respondent, thus permitting an assessment of job satisfaction.

Other questions focussed on worker views on automation and its consequences, on worker participation in decision making, and on worker expectations of his union and his estimates of union performance. Another group of questions dealt with social issues outside the plant. Workers were asked to indicate the degree of their interest in such issues and their involvement in activities designed to affect such questions.

It was decided that a sample of workers in automated and non-automated units should be interviewed.

To assist the standardisation of the data collection procedures, a Field Work Manualwas prepared which served as a guide for each research team. In Part I the manual contained suggestions for the kinds of background material on the project which might be made available to the research staff and to management and trade union officials to help them understand the project, and it emphasised the importance of having each national team review national literature on topics covered by the study.

It re-stated the formal recommendations adopted at project meetings that workplace observers should be persons who were not interviewing workers and who had experience with measurement instruments. For worker interviewers it was urged that they be given training in the technical aspects of the project. Such training was to help interviewers become "familiar with the terms used and with the meaning and intent of specific items of the questionnaire". As a further effort to avoid terminological misunderstandings, definitions of twenty-seven key terms were provided.

The manual also provided guidance on the sequence and timing of field operations and in Part II explained how specific points in the research instruments were to be handled. In Part III the handbook dealt with procedures for selecting plants, units, workplaces, and workers to be studied. It noted, for example, that part-time workers and non-permanent residents should be excluded from the project sample, and suggested procedures to follow if it were necessary to select a sample of a unit's work force for interviewing purposes.

In effect, the choice of plants was undertaken according to certain guidelines in order to reduce inter-plant differences to a minimum. Work units were similarly chosen in accordance with established criteria. This is not to claim that the study is based on a sample which is representative of the country's entire work force, or even of all automotive workers. Rather, since the sample was chosen in a purposeful manner, according to certain objectives, it can be seen as representative of a segment of the automobile industry.

At the same time, there are certain problems of methodology which may have had some influence on the findings. In some languages, for example, the translation of words cannot yield the same nuances as are contained in the original phraseology. There were necessarily some differences in the selection of job categories, and different cultural patterns had an unavoidable influence on a few of the questions[3]. It will be in the interpretation of cross-country similarities and differences that these points will require consideration.

A separate book of coding instructions was also developed.

4. COORDINATING ARRANGEMENTS

In order to have a formal structure for the project which would effectively co-ordinate the interests of the various parties involved and integrate it into the existing infrastructure of the Vienna Centre, a "Steering Committee" was established. This group was composed of one person from each participating country and two representatives from trade unions (East and West), together with an official of the International Labour Office who had been specialising on problems of automation.The President of the Board of Directors of the Vienna Centre and the Scientific Secretary were ex officio members, and their participation in all meetings of the committee was significant for the success of the project.

The committee meetings were informal and consensus was always an objective. "Overnight discussions" led to agreement on all but the most intransigent issues; there were only a very few occasions when voting was necessary. As the body formally responsible for the project, it dealt with policy questions arising from the research as well as the basic scientific and conceptual issues. In 1975 it was found advisable to establish a small Executive Committee to act between meetings of the Steering Committee with its decisions to be subject to review by the parent body.

Working group meetings were organised to deal with specific questions of a technical character. They normally brought together about twenty participants, with variations depending on the availability of the researchers and the location of the meeting.

Each national team financed its national research. In addition, a number of teams provided contributions to the common expenses of the project: through direct financial contributions to the Vienna Centre for project expenses, the provision of computer time and services, or as host for various of the international plenary, working group, or technical group meetings.

Researchers in this project have been cognisant of the difficulties surrounding their work. Tracing cause and effect is always a complex endeavour: can the effects of automation be separated from effects arising from other changes - in work organisation, management behaviour, the social and economic environment? The papers in this volume present the first and partial view of what this project may contribute to an understanding and evaluation of automation's effect; the national reports for other countries and the international comparisons still to come will provide additional needed perspective.

NOTES

1 A review of literature was prepared specifically for the study: Julius Rezler, The Effects of Automation on Job Content: a review of relevant literature as contribution to the Moscow Conference of the international research project "Automation and Industrial Workers".

2 The texts are reproduced in Part III of this book.

3 For example, the question: Do any of your supervisors or the administrative staff invite you to their homes? was accepted as a question even though it was known that it had different shades of meaning according to general country practice about inviting acquaintances - of what social level - to one's home.

3. The Experience of the International Comparative Study

V. USSENIN AND M. NOTCHEVNIK

1. APPROACH TO INTERNATIONAL SCIENTIFIC CO-OPERATION

The atmosphere of peaceful coexistence is conducive to the establishment of scientific contacts and to scientific co-operation in essential problems concerning the interests of people of many countries. It is natural that co-operation among scientists from countries with different social systems should have developed first in the technical and natural sciences. Joint scientific research in the social sciences is much more complicated because of the different ideological positions of participants in these studies. Nevertheless, it is desirable to undertake such research because of the value of revealing tendencies in social and economic development and of comparing different national experiences in solving specific problems.

In the Final Act of the Conference on Security and Co-operation in Europe, signed in Helsinki in August 1975, the participating states declared their intention to contribute to international scientific co-operation. As one of the fields of such co-operation, the Final Act names the humanities and social sciences including "comparative studies on social, socio-economic and cultural phenomena".[1.] At this conference, Secretary General of the Central Committee of the Communist Party of the Union of the Soviet Socialist Republics, Leonid Brezhnev, stated: "Possibilities for co-operation extend now also to areas where it was unthinkable in the years of the cold war".[2.]

There are at least three kinds of international comparative study: unilateral, centralised, and multilateral.

1.1. Unilateral International Comparative Studies

These studies are carried out by scientists in one country studying some social problem on the basis of materials from two or more countries. In this case, scientists unilaterally choose the literature to be analysed, make comparisons, and draw general conclusions.

Studies of this kind are carried out in many nations not only for theoretical but for educational and practical purposes.

International comparisons of social phenomena and trends of social development made in reports of many international organisations, including the UNO specialised institutions, also belong to the kind of unilateral comparative studies. These publications are valuable sources of information, but they are not generally based on the collection and analysis of primary materials. Rather, they summarise official data or the conclusions of separate studies undertaken at different times, for the different purposes, under non-comparable conditions and by different methods.

Unilateral international comparative studies are not the result of international scientific co-operation, and in this respect they distinctly differ both from centralised and multilateral international research projects.

1.2. Centralised International Comparative Projects

These projects provide for co-operation of scientists from different countries and are characterised by the following features:

(a) The organiser of the whole research is one central (international or national) scientific institution which determines the orientation, scope, forms and methods of work, and formulates ideas and concepts. Further, it sums up the results and takes upon itself the responsibility for the content and conclusions of the work done. An international project of this kind is often guided by one scientist, and the final results are often published under his name.

(b) The forms of participation of foreign scientists in such international projects differ. Sometimes they are assigned only the role of experts, consultants, or participants at meetings and symposia organised by the central scientific institution. In the resulting publications their participation is usually marked by gratitude only. In other cases the foreign scientists (or foreign research institutions) are only entrusted with the study of a certain problem within the country they represent. Their contributions are strictly national (not comparative) though published within the framework of the general project under their own name.

1.3. Multilateral International Comparative Studies

These studies represent the closest form of co-operation of scientists from different countries on the basis of equal rights. Essential to the success of this form of research are: the working out of agreed goals, the choice of comparable objects for investigation with due regard to particular conditions in different countries, adoption of a common programme and methods of research, and a serious scientific interpretation of the data obtained.

It is of especial importance to work out research instruments so that they will reveal accurately the special features of the phenomena under investigation in different countries rather than concealing these peculiarities or even misreprenting the real situation. Therefore, research instruments for international comparative studies cannot be borrowed mechanically from the store of instruments available for research on similar problems in a single country. It is necessary to adapt them and even to create new ones for use under conditions prevailing in different countries, especially if they do not have similar social systems.

Such studies lead to many difficulties. It is often necessary to change conventional methods. Resources at the disposal of research workers vary in different countries, also the issues of primary interest. Overcoming these difficulties implies a high level of responsibility on the part of research participants, a readiness to regard carefully the positions of their counterparts, and a sincere aspiration for scientific objectivity.

Assuming that these problems can be answered, such studies can be expected to lead to substantial improvement in the quality of social research.

The "Automation and Industrial Workers" project is an attempt to apply this model. This model for co-operation was developed within the framework of the twenty or so joint research projects initiated and co-ordinated by the Vienna Centre during its 14 years of existence.

In these projects, the Vienna Centre has acted as an active intermediary in the work of scientists of different countries, not as a centralised institution directing the research. All participants in these projects have equal rights, and jointly determine the goals and the way in which the work is done.

In evolving this model the Time Budget study previously launched by the Vienna Centre was a significant step. This study examined the different uses of time by city dwellers in twelve countries (7 in Eastern Europe, 3 in Western Europe plus the United States and Peru). The results of this research as well as a thorough analysis of its organisational approaches, were published in The Use of Time.[3] Alexander Szalai, Director of the project, wrote in the introduction:

> "A truly multi-national organisation of a cross-national comparative research project presupposes that the basic research design is developed collectively by all participants in the project, all standards of survey procedure (choice of survey sites, sampling, interviewing, coding) be worked out and agreed upon in common, and the whole body of collected data be made accessible to all for evaluation and interpretation giving ample opportunity, on the one hand, the elaboration of common views and ideas of the participants about the results achieved, and, on the other hand, for the validation of their individual approaches and judgements."[4]

The completion of the Time Budget study coincided with the beginning of the still more extensive international comparative research project co-ordinated by the Vienna Centre, "Automation and Industrial Workers". Despite the common origin, important differences have emerged in the conduct of these two projects.[5]

Firstly, the nature of the problems investigated in the "Automation" project required collection of data from a wider variety of sources. The Time Budget study was based on personal interviews with citizens, whereas the "Automation" project necessitated a survey of the opinions of industrial workers, expert evaluation of work processes, information from trade union and plant management, and a review of main characteristics of the national socio-economic system. Therefore, quite different research instruments had to be created.

Secondly, the participants of the "Automation" project gave higher priority to the principle of equality of all national research groups. In particular, instead of the appointment of one director of the project, as in the case of the Time Budget study, or two co-directors representing socialist and non-socialist countries, it was decided to form a Steering Committee consisting of representatives of all national research teams. This enhanced the collective nature of the project administration.

Thirdly, not only scientific workers but trade union and management representatives took an active part in the organisation of the project. Two representatives of trade unions were included on the Steering Committee. Participation by the trade unions and management contributed a valuable perspective.

Finally, despite different ideological views, participants in the "Automation" project developed a compatible collaboration in the pursuit of their common scientific interests.

The representatives of different fields of social sciences and trade unions from fifteen countries were not always interested in the same aspects of the investigation. But, as shown below, the group was able jointly to define a common research programme and main hypothesis.

2. THE STUDY OF THE SOCIAL CONSEQUENCES OF AUTOMATION

One of the contemporary problems that attracts the ever-growing and constant attention of working people, trade unions, and the general public is the social consequences of the scientific and technological revolution of which automation is a part.

This development affects different aspects of life in modern society. The influence is felt, first of all, in the economic sphere. The scientific and technological revolution creates opportunities to remake radically production methods, to create basically different and highly productive work tools, completely new materials, and, as a consequence, new branches of industry.

The broad introduction of the latest scientific and technological achievements in production results in changes in relations between man and machine, and in relations between the people not only within the enterprise but outside as well. In this connection, different social classes, government bodies, political parties, trade unions, and other organisations strive to influence the pace of scientific and technological progress and its social consequences.

In almost all countries, many publications are devoted to revealing these consequences, to their assessment and prognosis, and to the working out of measures to meet them.

The investigations undertaken by scientists in socialist countries point convincingly to the fact that scientific and technological progress under socialism has become one of the main tools for raising substantially the material and cultural living standard.

The possibilities in a planned economy permit conscious direction of the pace of introduction of automation and other forms of scientific and technological progress. Nevertheless, the mere existence of the socialist system does not eliminate automatically all problems and difficulties related to the scientific and technological revolution.

Overcoming these difficulties implies great efforts and big organisation on the part of public and social bodies. In his Report to the 24th Congress of the Communist Party of the Soviet Union, Secretary General of the Party, Leonid Brezhnev, pointed out the current task of historic importance - that is "to unite organically the achievements of scientific and technological revolution with the advantages of the socialist economic system". The vitality of this task was underlined also at the 25th Congress of the CPSU in 1976.

A characteristic feature of numerous works by Western authors dealing with social consequences of automation and other forms of scientific and technological progress contain opposite evaluations of existing phenomena and conflicting prospects for their development. A number of authors represent automation and all new types of technology as a disaster and curse for mankind. There are opinions that technology becomes an autonomous force, that it becomes uncontrollable by man and destroys human values. Other authors, on the contrary, praise to the sky the possibilities of technology and believe it to be the automatic solution of all social problems.

The contradictions in evaluation of social consequences of scientific and technological progress existing in works of different Western authors are explained to some extent by the specific orientations of investigators, and by the different materials which they use. The contradictory conclusions of different authors concerning social consequences of automation are also explained in a number of cases by the fact that different levels of technology are chosen for analysis. The term "automation" is very widely used and often refers to very different levels of technology.

But no doubt these different evaluations reflect also the real contradictions of scientific and technological progress, which could serve for common good, but lead very often to negative social consequences especially for industrial workers.

But not all technological innovations exert a similar influence on workers. For example, although there is a general tendency to raise the qualification requirements of workers, some types of automated equipment can, on the contrary, be operated by workers having a lower qualification level.

It is difficult to separate the technological factor from others that influence social consequences of automation. As pointed out in a report on this topic prepared by the International Labour Office to the Conference in 1972:

"Existing statistical series do not usually enable the influence of technology to be identified among the many factors determining social conditions. Where research has been undertaken with the specific purpose of isolating the technology factor and of studying its effects, these studies refer to individual cases in a particular and often unique setting, and the results therefore rarely lend themselves to generalisation. The findings of attempts to forecast future developments, even if over relatively short time spans, are even more divergent because such forecasts involve a departure from the solid basis of existing facts and enter the realm of imagination and of more or less enlightened guesswork".[6]

In many national and sometimes international studies, certain phenomena of social life are attributed to the exclusive effect of new technology. The extrapolation of these phenomena in direct proportion to the introduction of technological innovations, and without due regard to social mechanisms, often leads to paradoxical results. For example, a number of social phenomena considered to be a direct consequence of automation in some countries are wholly absent in other countries having a similar level of technology.

How is it possible to distinguish among the number of factors the influence of new technology upon work and life of people? One approach is to try to separate the influence of technology from the influence of the social and economic system of a country.

One cannot consider unemployment, underemployment, sharp fluctuations of the economic cycle, and shut-down of enterprises as natural or even inevitable results of the introduction of automation, if these phenomena are characteristic only for some industrialised countries. Such phenomena are connected only with specific economic and social conditions. The introduction of automation can only influence the extent of these phenomena. But though this fact seems to be obvious, one very often disregards it in international comparative works. In a number of cases the conclusions applicable only to some countries are automatically extended to other countries as well.

Along with national studies, an international comparative research can be of significant interest, especially if it is carried out in a number of countries with different economic and social systems and the investigation is conducted according to a common programme and by common methods.

3. THE PRACTICE OF CO-OPERATIVE INQUIRY

In spite of sharp disagreement on many issues, the participants in the Automation project displayed constant aspiration for scientific co-operation and for achieving constructive results. This enabled them to overcome a succession of difficult problems of conceptualisation, methodology, and particularly of securing comparability in the research across widely different systems and cultures.

One of the principal questions to be solved concerned the role of the social and economic system of the country. Different opinions were expressed by the participants on this topic. Some of them argued that the social system of the country did not influence the social consequences of automation. Automation was to be regarded as a quite independent phenomenon.

The others were of the opinion that the social system of the country influenced not only processes of automation but had also a direct influence on social consequences of automation and on the workers' attitudes to work and life.

After a long discussion it was agreed to incorporate in the research design the following statement as one of the main hypotheses:

> "The political and socio-economic system of a country exerts an essential influence both on the processes of introduction of automation and on its social consequences."

To test this hypothesis it was necessary to do additional work, i.e. to study the ways and methods used by society to influence the development of automation and its social consequences. It was necessary to give a characteristic of the social system of the country underlining those elements in the system which were of importance for the study.

It was very difficult to come to an agreement on this point. Some participants objected to the idea of additional research work saying that it was not necessary, that all the data of interest to us could be easily taken from existing reference books. The others objected to the idea because of an opposite argument. They said that it was very difficult to find and summarise necessary facts and information.

Many other participants insisted, on the other hand, on investigation of social factors which influenced or at least could influence social consequences of automation and worker attitudes to this particular phenomenon. This position was based on two main arguments. Firstly, if the research showed that there were different social consequences of automation in different countries, it would be necessary all the same to give later some valid explanations about this fact according to the nature of social factors.

Secondly, irrespective of the future results of the study it was necessary for the purpose of comparison to know the character, nature, and meaning of the social institutions one would meet in each country. In some countries there are special institutions which do not exist in other countries. Take, for example, the socialist emulation system, which does not exist in Western countries. In the Soviet Union it is impossible to show the workers' attitudes to work and life without considering this factor. But the meaning of it cannot be known properly in Western countries.

Another, maybe even more important, factor is that in different countries many social institutions have the same names but quite a different meaning. Take, for example, "workers' participation in decision making." In order to compare such an institution it is necessary to have a detailed explanation of its meaning and functioning for each country.

As a result of a very prolonged and hard discussion the participants of the study decided to include in the study (as a special part) the description of the social system of the country. It was meant, according to the standard form, to give an explanation of different ways and methods used by society to influence technological innovations and their consequences. The role of state, political parties, and trade unions, was to be considered in this respect. The most important features of existing labour legislation and social security system had also to be given for each country. This part of the study will help a good deal in the future evaluation of research results at the enterprise level.

To conduct the investigation it was of principal importance to work out appropriate research instruments. In the course of joint and very intensive work, five research instruments were prepared. The first of them contains the detailed instructions for the description of a social and economic system of a country. The instructions specify those features of social systems that are of most importance to the study, the elements of power structure and of economic development process, job, and income security system existing for the workers. The second research instrument contains the text of an interview with management of the enterprises under investigation. The third instrument is the text of an interview with trade union representatives at national and plant level. The fourth instrument contains the instructions for carrying out workplace observations in automated and non-automated units. And the fifth and last document is the text of the questionnaire for industrial workers. This questionnaire was to be used by means of personal interviews.

Each of the above-mentioned instruments was important in itself, but their content was interrelated and should make it possible to reveal different factors.

Specific scientific problems arose in all sections of the research instruments because of their intended application to many countries with different social systems, different social, and cultural habits and ortientations. Most of these problems concerned the questionnaire for industrial workers. The experience of working on this text could be of importance for the practice of international comparisons.

The purpose of the questionnaire was to reveal both objective changes in the position of workers in the process of technological innovations and subjective perception of these changes by individual persons, new aspects in work motivations, and in social orientations of workers.

As the same questions were to be addressed to workers employed in many countries, the very content of the questions and their formulation required thorough consideration in order to ensure correct understanding, to provide comparability of answers, and not to veil principal peculiarities in the positions of workers in different countries.

Some differences in understanding questions could inevitably arise because of a specific character of social-psychological atmosphere.

The variant of the questionnaire for workers worked out at the first stage revealed a number of significant shortcomings during the pre-test study. In particular, the questionnaire turned out to be too long. This made it difficult to receive precise answers to all questions from the workers.

A number of questions taken from questionnaires used in previous national studies turned out to be quite unacceptable and alien for workers in some other countries, for instance in the Soviet Union and in Finland.

Some groups of questions began with the statement: "Some people expressed the following opinions. Do you agree with them?"

Many Soviet workers answered that they not only disagreed with the opinions, but such opinions did not exist at all among them. A number of questions from this variant of the questionnaire devoted to social orientations were particularly connected with money and ways of making money. The obtrusive repetition of such questions aroused perplexity and even irritation among workers at Soviet enterprises. When workers were asked to express their opinions on some of the statements they often burst into laughter, and serious attitudes to the interview disappeared.

These are some examples of such statements: "To make money there are no right or wrong ways any more - only easy and hard ways"; "Most people really don't care what happens to the next fellow"; "It's hardly fair to bring children into the world with the way things look for the future "; and so on.

These and a number of other methodological shortcomings of the questionnaire required its substantial remaking. This task proved to be far from simple, and only as a result of prolonged discussions were the ways of solving many problems found.

The new text of the questionnaire turned out to be more acceptable. However, it also involved certain difficulties for workers at Soviet enterprises. So, for instance, workers had difficulties in giving a concrete answer to the question of how important to them was the opinion of "their religious (church) leaders". Even the proposed alternative of answer that this opinion was of no importance could make an impression that the worker still had such a religious leader and only his opinion did not matter. Meanwhile many Soviet workers had little or no idea at all about the existence of such "religious leaders".

When we worked on the elaboration of research instruments we had also to overcome many difficulties connected with different general approaches to many concepts in different countries. One of our tasks was to investigate whether the workers participated in management and whether the introduction of automation influenced this kind of participation. Some questions had been included on this topic.
But it was found out rather late that the notion of participation in managment was very different in different countries and the answers given by the workers were practically incomparable. It was necessary for the purpose of this particular study to give a specific

definition of what was to be understood by worker participation in management and decisionmaking. Such a definition was made. But it was not enough. It was necessary to reformulate all the questions on this topic in such a way that they might be understood in the same meaning by workers in different countries.

Another example can be taken. In order to study creativity of the workers it was decided to investigate whether they participated in a suggestion system, how many suggestions were made in previous years in automated and non-automated units. We have got such data concerning the differences among participating countries. But when we began to analyse and interpret these differences it was found that they were due not only to the really different number of suggestions made by workers, but also because of a different interpretation of the term "suggestion". In the Soviet Union, for instance, there is a very developed but rather formalised system of rationalisation and innovation suggestions. These suggestions are to be made in a written form; they must correspond to certain requirements. The workers also make (and very actively) different proposals for improvement of technology, methods of production at the workers and trade union meetings, at union committees, standing production conferences, and so on. The management is obliged to take these proposals into consideration and report about their implementation at the workers' meetings. But this kind of less formal proposal is not regarded as an official suggestion.

During our evaluation of the pre-test data it was found that in some other countries there is a quite different approach to the same problem. Every formal or informal proposal made by the worker in connection with improvement of technology or methods of production is regarded as a rationalisation proposal. This proposal can be made informally to the foreman, supervisor, any manager, or can be made in any other way, but it will be regarded as a rationalisation proposal.

Striving to secure comparability, the participants of the project came to the common conclusion that for the purpose of our study we should regard as a suggestion in all countries not only formal, i.e. relatively big innovations, but also informal proposals made by the workers and designed to introduce small improvements.

However, this decision was afterwards nct always possible to implement. The workers at Soviet enterprises were so used to their own understanding of the term "suggestion" that they disregarded our interpretation and gave answers according to their comprehension. The official data about the suggestions that we could get at Soviet enterprises also did not correspond to the interpretation of the term we had agreed upon.

Our common sincere desire to secure comparability of the results of the investigation had produced another problem. Was it necessary at all to study and consider factors and phenomena that existed only in some countries and were absent in others? Some participants of the project argued, for instance, that it was senseless to ask the workers about their participation in economic planning when it was already known that such kinds of workers' participation did not exist in a particular country. The answer to this question received at the workplaces in socialist countries would not allow comparison with other countries. So it was suggested to omit this question altogether.

The same position was taken in respect of many other questions. Was it necessary to show in the study the role of political party organisation at workplaces, the role and forms of a socialist emulation system, when such kinds of institutions existed only at workplaces in socialist countries? It was said that there was no field for comparison here.

Other project participants could not agree with these arguments. The international research had to study the real situation in each country and the comparison had to reveal common and specific features in workers behaviour, attitudes, and in an objective situation. For the purpose of simplification one could drop, of course, many things and create an artificial field for comparison of equal factors. But the real situation at socialist workplaces was impossible to understand and evaluate without considering the workers' participation in economic planning, the role of a plant political party organisation, the role of a socialist emulation system, and so on.

At workplaces in Western countries one could also find phenomena that did not exist in socialist countries. Take, for example, pluralism of trade union organisations. In many Western countries not one, but several trade unions operated at the same enterprise. It would be unwise to suggest disregarding this fact only on the ground that in socialist countries only one trade union organisation exists at each workplace.

After a long discussion a solution was found which was quite acceptable to everybody. It was agreed that in interviews addressed to management and trade union bodies all the questions would be put in a standard form. The managers and union leaders determine themselves what is applicable and inapplicable to the given situation.

It was also agreed that it was not desirable to put to the workers such questions the meaning of which might be unknown to them. Each national team was to decide if it was possible or not to put to the workers questions concerning economic planning, social planning, and so on. In respect to the workers' participation in management, standard lists were elaborated. They enumerated different possible spheres of worker participation in management and possible ways and methods of such participation. Each national team was to decide on the ground of the given situations which items were to be dropped or to be added in order to make the list understandable to an ordinary worker. The Steering Committee was to be informed about all these changes.

Some questions were to be addressed only to workers of a group of countries when it was clear that these questions concerned only this particular group. The example of it was different forms of socialist emulation systems.

At that stage of the work principal decisions were reached; in order to implement them a special group of experts was created and authorised to finalise all research instruments. This group worked during January 1974 in Poland.

It was not easy to take into account numerous suggestions from fifteen national teams, especially when some suggestions contradicted others. But the work was still done.

The joint work of scientists from countries with different socio-economic systems on undertaking international comparative projects in the field of social problems is still a rare case. The prospect of expanding international scientific co-operation requires taking into account the acquired experience of such work, no matter how modest it is.

It is true that studying the experience of others cannot eliminate the necessity of gaining one's own experience. Even coming across the analogous methodological problems, new collectives of scientists in each concrete situation must search for their own ways of solving these problems. Nevertheless, this fact does not reduce the importance of previous experiments which proved the feasibility of undertaking international scientific research projects, projects which seemed almost unrealisable before and which showed an example of possible ways of solving the complicated methodological questions of international comparison.

REFERENCES

1. Conference on Security and Cooperation in Europe, Final Act, Helsinki, 1975, p. 130.

2. Conference on Security and Cooperation in Europe (Stage III, Helsinki, 30 July to 1 August 1975). Verbatim Record and Documents, p. 63.

3. Alexander Szalai (ed.), The Use of Time, a publication of the European Coordination Centre for Research and Documentation in Social Sciences, The Hague, 1972 (868 pp.).

4. Ibid., p. 11.

5. For a comprehensive discussion of these problems, see in particular, Cross-National Comparative Survey Research: Theory and Practice (edited by A. Szalai and R. Petrella in collaboration with S. Rokkan and E. K. Scheuch), Pergamon Press, Oxford, 1977 (498 pp.).

6. Labour and Social Implications of Automation and other Technological Developments, ILO, Geneva, 1972, pp. 1-2.

PART II

1. Automation and Work Organisation—A Swedish Experience*

JAN FORSLIN, JENS SODERLUND, AND BARBRO ZACKRISSON

1. BACKGROUND

The Swedish part of the Automation project can be seen as a step in an on-going social research process about working conditions, particularly research in connection with technological change. The aim of the study is to point out negative unintended social consequences and risks as well as to utilise inherent possibilities. It is hoped that the relative perspective given by an international comparison will also contribute to the understanding of the social effects of technological change in Sweden. But the national study has a value per se as a direct investigation of Swedish conditions.

There is a strong link between production technology and human working conditions. Apart from the management philosophies regulating the working conditions, the production technology, e.g. level of automation, has been attributed a vital role in determining not only the physical but also psychological aspects of work.

* Financial resources for the Swedish participation have been provided by the Swedish Council for Personnel Administration and the Swedish Work Environment Fund.

The project is supported by the Swedish organisations involved in the basic problem under study.

The Volvo Skovde plant has generously given access to information and provided facilities for the fieldwork. Union and management representatives have contributed in terms of ideas and knowledge.

The empirical work has, apart from the authors, been carried out by Viveka Ferlander, Ann-Marie Nobel, and Ulf Svensson. The project has benefited also from their assistance in data interpretation and their participation in the more general discussions on main issues of the research.

Often there have been great hopes that technological innovations would free man from heavy and risky jobs - improved physical working conditions. At the same time there have been fears that the jobs would be unqualified and routine - impoverished psychological working conditions.

In general there has been assumed and to some extent verified a U-shaped relationship between the level of mechanisation and job satisfaction. Thus mechanisation lowered the relatively high job satisfaction prevalent in the pre-industrial era. And now the hope is that the next technological step - i.e. automation - will again improve the workers' satisfaction.

This assumption can be said to be technologically deterministic. It implies that technology is neutral and decides working conditions without any possibilities for social influence.

The industrialisation of production life and mechanised mass production in particular have had an immense impact on human working conditions. There has been in Sweden, as in many other countries, a systematic transition from qualified artisanry to unskilled routine jobs at the same time as the industrial labour market has expanded. The philosophical innovation behind this development was the scientific management school of thought (Taylor, 1910) which was widely applied also in Sweden. One implication of this philosophy was the separation of the intellectual parts of the work from the manual; reducing the worker to a physiological machine, anonymous and replaceable - lacking any human quality.

The principles of scientific management were partly developed within the automotive industry and were given their most thorough-going application there. In recent years the strongest opposition against these principles is also to be found in this branch of industry in Sweden.

The organisation principle was to transfer the qualified part of the work to specialist functions like departments for repair and maintenance, tooling control and planning, or to the supervisory system. What then remained of the job for the worker was organised to demand as little training as possible, the tasks being repetitive, with extremely short work cycles, uninteresting, giving few opportunities for development, learning, initiative, or responsibility.

The scientific management philosophy should be seen with its historical background. During the period of rapid industrialisation there was a need to transfer labour with low education and no industrial experience from agriculture to industry.

But the philosophy also reflects an image of man and assumptions of human nature and motivation: essentially a "stick and carrot" assumption. The prime motivation to work is seen as monetary reward; the worker has no interest in the work per se nor any wish to take responsibility and develop his personal abilities. He has to be closely controlled and supervised in order to do his work. When such assumptions are made the leading principles for work organisation and divison of labour they become self-fulfilling because work not only produces goods but also shapes the individual at work.

There is no need here to recapitulate th criticism of scientific management, but it should be said that in spite of the many negative consequences from its application it also meant great progress for production and also for the worker. It should also be emphasised that the main thinking of this ideology which developed during the nineteenth century still dominates modern working life not only as the principle for production planning but also the basis for the training of generations of management and technicians.

Starting with Marx, numerous sociologists have formulated very sharp criticisms of industrial working conditions. In the 1950s many now classical and internationally well-known studies were produced, e.g. Walker and Guest (1952) and Chinoy (1955), which clearly pointed out negative social consequences of industrial work in the automotive industry. Later, modern literature on alienation gave this criticism a more psychological content, e.g. Seeman (1959), Blauner (1964), and Kornhauser (1965).

During the 1960s new theories of human motivation and alternative images of man gained acceptance and widespread application in the training of management and specialists in industry. Abraham Maslow (1955), Frederich Herzberg (1966), and Douglas McGregor (1960) are the most important contributors. Essentially the idea was that man has higher needs that he wants to have fulfilled in work that determine work motivation and work behaviour. Man does not even primarily want pay out of work but has needs for self-actualisation.

During the 1950s and 1960s the Tavistock Institute in London had shown the possibilities of a different kind of research than traditional descriptive sociology. This research is more preoccupied with active participation in social change. The research is a means of bringing about change within social systems, e.g. within a work organisation.

Also from the Tavistock Institute came the view on production units as socio-technical systems (e.g. Herbst, 1974), the implication being that two subsystems - a social and a technical - in a production system have different "needs" and different qualifications, the social system being more suitable for problem solving and creativity, while the technical system is better equipped for physical strength, repetition, and precision. An important concept is that of organisational choice, which is in opposition to the long dominant idea of technological determinism in connection with work organisation. By organisational choice is meant that, given a certain technology, there are many different ways of organising the work, each of which will give different work situations and optimising different criteria.

From the view of alternative work organisation it follows that it is almost impossible or even unnecessary to make assumptions about relationships between technology and job satisfaction because the social effects come from the organisation principle rather than from the technology itself. Then it becomes a question of values rather than a "neutral" technical question. However, it should be said that different technologies have possibly different numbers of degrees of freedom in the choice of organisation.

Yet a more radical approach would be to state the desirable social effects as well as the technical and economic requirements and then to develop the technology and its application to meet also the demands from the social system. For ergonomic and medical aspects of work this is now done to different degrees, e.g. by the use of threshold values for permitted air pollution or noise. For psychological needs a similar procedure is not yet in practice.

At present there is a strong tendency, at least in Scandinavia and also elsewhere, to seriously question the principles of traditional production organisation. Great efforts are made to find new solutions. This phenomenon should be seen against the background of the influence from the social sciences as well as current changes in society as a whole, in levels of education, in new values, and new aspirations.

* * * * *

This report has been regarded as an opportunity to call attention to problems and observations in connection with automation that are considered important in the national context. In the Swedish case the question of work organisation and technological change is a central aspect of the present development in working life and has thus been given high priority here.

When the investigation was carried out one of the new transfer line departments where a new work organisation had been applied had operated for about a year. This department was included in the sample, thereby giving a unique opportunity to shed light on the moderating effect of organization on technology. For this reason the Swedish national report has mainly been arranged as a comparison between the new unit and the traditional conventionally organised units.

This approach is also justified by the observed lack of variation in working conditions in connection with different technology in the traditional departments. This has been documented in a previous report (Nobel and Svensson, 1975).

As the new unit only contains about twenty workers, a word of caution should be stated about the interpretation of the results. The statistics are based on very few persons and should, therefore, be handled with care. The general overall tendency should be interpreted rather than specific percentages.

2. RESEARCH SITE

2.1. Description of the Plant

As the investigation was directed towards the automotive industry there were only two possible factories to be chosen in Sweden. One of these declined the invitation due to work load. One factory remained which also is the largest of the two - Volvo-Skovdeverken. The next task was to find departments with suitable technology. Five departments were possible from this point of view, having the advantage of being located in the same plant.

The plant produces petrol and diesel engines. The number of employees is about 4500 of whom 3800 are workers. About one-third of these are immigrants, most of whom come from Finland. The company is located in a town, Skovde, with about 35000 inhabitants, situated in the south-west of Sweden.

The people recruited are, with few exceptions, young, and there has been a policy during recent years to recruit larger percentage of women. Today 18% of the work force are female, that target being 30%.

In the company there is a special group which continuously follows issues concerning working environment. In that group the project idea was introduced and accepted. In this way both the local management and the trade union had their say. In both cases there was an interest shown in the investigation as guidance for future work in developing working conditions.

As the investigation was seen as a tool to improve working conditions, great emphasis was put on feedback of results to the plant. After a very quick data processing and a rough analysis of data, discussions of the results were arranged on different levels of the plant. This had several purposes:

(1) To meet the formal obligation of the team to report the results to departments involved as well as to the management and the union;
(2) To allow discussions so that participating respondents could contribute to the interpretation of the data, verify the interpretations made by the research team, and fill in gaps of knowledge;
(3) To make the investigation an impetus to action and change.

The results were used as a means of making the workers conscious of their situation as well as showing possible alternative ways of arranging their work situation. This was a way to contribute to their discussions about their working conditions.

All the employees of each participating department were gathered in groups of ten, where the main results and the particularities of their own department were discussed.

To the trade union, the local management, and the technical staff, the results were presented as an evaluation of the present working conditions and of the changes that had been undertaken. In this context the study had a great value as a fairly neutral evaluation made by investigators from the outside.

The results were used in a one-day seminar devoted to work organisation and addressing all management of the plant. At this meeting the experiences that had been made in different departments, where new models had been tried, were presented to be used as an impetus to further activities in new directions. The evaluation included economic, technical, as well as job satisfaction aspects. Here the results from the automation study made a very good contribution, being an aggregated, formalised, and less subjective account of the opinions of the employees.

2.2. Description of Technology

The complex and composite nature of engine blocks necessitates a flow line production. This is an adaptation of flow process technology to discrete items (Wild, 1975). The units where transfer machines are in use represent transfer flow lines. The newest department though can be distinguished as a link line, since the intermediary transportation steps are automatic as well as the machining.

All automated equipment consists of transfer lines, while non-automated equipment consists of single-purpose machines, e.g. radial drilling machines. Principally, in all investigated units there is the same type of equipment but the proportion of automated parts varies. Figure 1 shows the proportion of automated operations in the units.

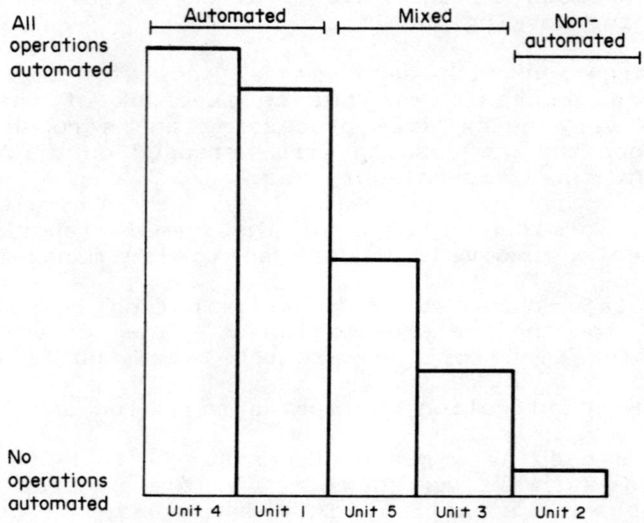

Figure 1. Proportion of automated operations in the units.

From the occurrence of different kinds of equipment it could be seen that automation does not necessarily mean handling of more advanced machinery. For example, manually handled equipment such as hand - measuring instruments and hammers, etc., are used by practically all workers. Control panels are used even less in the automated unit. The less direct operation of the machinery is illustrated by the much less frequent manipulating of control devices.(On pressing the results somewhat one can almost say that automation in this case means a return to a handicraft type of work.)

The principal steps of the production in all units follow approximately this sequence:
(1) Rough machining of surfaces (broaching and milling).
(2) Rough machining of holes (drilling and tapping).
(3) Fine machining of surfaces (broaching and milling).
(4) Fine machining of holes (honing).
(5) Washing, control and adjustement.

(In some of the departments the initial step is painting of blocks.)

The following five departments were chosen to represent different levels of automation:

Unit 3: Producing blocks for diesel engines. Very low level of automation. Only single-purpose machining. Equipment consisting of mills, drills, reamers, tapping machines, recessing machines, and washing tanks. In this unit there were 12 operators working and 1 setter. The unit had no first line supervisor and operated in one shift.

Unit 2: Producing cylinder heads. This unit had a very limited amount of mechanised equipment consisting of one short transfer line, the rest: testing tank, mills, drills, broaches, washing machines, blower vibrator, and drill reamer. The unit operated in two shifts, each with 28 operators, 1 setter and 1 "other". The unit has one forman each shift plus one supervisor.

Unit 5: Producing blocks for diesel engines. As with unit 2, this is a department with a mix of machinery and a major part of production in transfer line. Equipment consisting of: mills, drills, recesses, bores, mill broaches, broaches, radial drills, and washing machines. Each of two shifts consisted of 16 operators, 3 setters, 1 "other", and 1 supervisor. Only one shift was included in the study.

Unit 1: Producing blocks for petrol engines. A mechanised department with one all-encompassing transfer line. According to the definition of the study this is an automated unit. Equipment consists of: painting box, drying oven, broaches, drills, mills, tappers, washing machines, measuring machine, arbor, and honing machine. In each of two shifts 24 operators, 1 both operator set-up man and 1 supervisor were working.

Unit 4: Producing blocks for petrol engines. Automated department: all operations performed by a transfer line. Input/output and transportation between transfer units are also automated. A few non-automated steps remain such as putting on bearing caps. The unit operated in one shift with 22 operators, 1 maintenance man and 2 supervisors.

In the following, units 1, 2, 3, and 5 will be referred to as traditional units and unit 4 as a new one.

2.3. Work Organisation

The new department in comparison with all the other units shows many differences regarding the structure of work organisation. The new technology in the department required and had made possible a new work organisation.
As was pointed out in one of the first socio-technical studies in the analysis of automation in a loom shed by Rice (1953), automated processes are continous rather than cyclic. This meant that efforts are concentrated on maintaining a steady state of production. This requirement breaks up the traditional work roles.

It would, however, have been possible to apply the traditional organisational principles with "one man serving one machine", but instead the group principle was agreed upon by the workers and management as a basis of the organisation.

Within the groups the members have shared responsibility for keeping the machinery running and they have been given much freedom to organise their own work. Quality control is also - to a considerable degree - left to the worker's judgement and not - as a general rule - regulated by detailed control instructions.

Further the sharp boundaries between different work roles, which characterise the traditional organisation, have been abandoned. To give an example, inspectors are participating in adjustment work or the adjuster is doing inspection work depending on actual work load.

One goal is that every employee will be given the opportunity to learn all tasks within the whole department. Apart from the spontaneous learning taking place within the groups, there is a systematic planning for further training of every employee. In addition to on-the-job-training there have been formal courses on machine technique, tools, and quality control.

The workers are paid on an incentive basis. The wage is based upon final output from the whole department. This means that the workers are dependent for their income on each other's efforts in both shifts. This differs from the ordinary case within the company, where workers are paid individually. In practice income variations between wage periods are low. Full pay is received when output corresponds to agreed achievement standards and delivery programmes. It is not possible for the workers to increase pay by producing more blocks than are demanded.

A summary of organisational characteristics of both traditional units and the new unit can be seen in Table 1.

2.4. Environmental Characteristics

The new unit differs in many aspects from the old departments, as in all other departments. The new factory consists of a main body containing the assembly and test departments, with four arms at right angles containing the machining departments. The latter are separated from one another by generous garden areas. This means that the social climate and environmental advantages of small workshops have been preserved despite the overall size of the factory.

Great efforts have been made to minimise negative working conditions such as noise and air contamination. Light and space are also better in the new department.

2.5. Sampling of Workers and Workplaces

The aim was to include all employees at each unit under investigation in order to get as complete a picture as possible. Further, the size of the units did not require a sampling procedure in order to reduce the number of interviews. The only restrictions in including a person were the rules laid down in the common instructions of fieldwork.

Table 1. Summary of organisational characteristics

	Traditional Units	New unit
Work instruction	Close and detailed	Only for guidance
Being tied to a workplace	Yes	No
Dependence on machine pace	Yes	No
Operator's tasks	Simple operations on machines	Process monitoring, maintenance, control of dimensions, exchange of tools and setting up machines, trouble shooting
Work role	Strict work role	Flexible work roles
Work cycles	Short, 1-5 minutes	Non-cyclical
Task assignment	One man-one machine	Group organisation
Wage system	Individual incentive system	Group incentive system
Specialisation	Operators highly specialised on simple tasks. Variations handled by specialist functions.	Broad competence in operators. Specialists work successively delegated to the operators through continous learning.
Communication	Strongly restricted possibilities to make contacts and because of noise and geographical restrictions	No geographical restrictions nor noise. Good possibilities for spontaneous contacts
Training	Very short period of training.No basic training the job and no increase in competence	Basic courses in production technique. Continous learning for on the job
Job rotation	No	Yes
Lay out	Line operation flow	Line operation flow

3. RESULTS

3.1. Functioning of the New Unit

In this section comments will be made on different aspects of the functioning of the department. When thinking of results one often tends to look for quantitative data showing output, absenteeism, etc. We are in this case not going to present any detailed description of figures and measures. Suffice it to say that budget standards are very well met and that the department - when compared to other departments - compares very well in respect to quality, production costs, and so forth. Labour turnover and absenteeism are extremely low for Swedish conditions. Absenteeism is, for example, below the figures noted for qualified maintenance personnel.

In this case, however, the qualitative aspects on how people behave and how the department functions are more important. It is also important not to look at the department as a closed system but to give special attention to what are considered as secondary effects, its effect on other systems and other system levels within the organisation as a whole. The conscious striving to humanise working conditions is not an isolated effort limited to some departments. On the contrary it is seen as a process of change involving the whole company.

3.2. Changes in the Work Content

As a general result we can point to the fact that almost all disadvantages connected with the traditional line technology have been overcome by the new technology and the work organisation. Such disadvantages have been

- monotonous and repetitive work
- lack of self-determination
- lack of learning and growth possibilities
- social isolation

Technical changes made are often of minor importance and have only marginal effects, as, for instance, when buffer stocks are introduced to make people less interdependent in the production chain. One should certainly not underrate the value of such minor changes, but it should be stressed that in the automation stage reached in the new engine block line the whole work structure has been radically changed. Simply put, the traditional work role could be described as being preoccupied with producing a number of parts within a certain limit of time. This is how the worker himself would define his work role. In the highly automated block line attention and effort are much more focused upon finding and eliminating technical disturbances in order to get optimal output from the total line. In this endeavour people are free to use their own judgement, to get together for discussions on technical matters, etc.

It is a general impression when observing and talking to people within the department that they are most of the time in an active and alert state both physically and mentally. This is different from what is often observed in departments with a conventional system. When, for instance, there is a work stoppage caused by some technical disturbance, people often sit down and the stoppage is experienced as a welcome break and an extra pause from the repetitive task. Technical disturbances are thus experienced either as a relief or as an irritating interruption in the work rhythm. Among people working in the new unit technical disturbances seem instead to be experienced as a challenge, something to learn from. When there is a stoppage workers gather together by the machinery looking for causes and discussing the problem. One-third of the workers also report that they have made suggestions during the last year as contrasted to 20% in the other units.

It has already been mentioned that the clear boundaries between different work roles have, to a very great extent, been eliminated as a result of the way work has been organised and that training has been given to the workers. This means that people act with much more flexibility than in traditionally organised production departments. When the supervisor and some of the workers were asked how the assignment of people to different tasks was decided, nobody could actually give an answer to that question. Normally people gathered before the shift in the rest area, and there it was informally decided who should work in which place. The greater flexibility also means that problems caused by changes in production rates, for instance, are solved much more easily than before.

As a consequence of the piece-rate system in mass production, workers are often found to have a short time-perspective. The individual worker sets as his personal goal to finish the day's work as soon as possible - normally much earlier than the formal working hours are ended. In contrast, the workers in the new unit look beyond the actual shift to outcomes over longer time periods. This in part has to do with the character of work content. For instance, one shift may lose production output as a result of stoppages caused by tool changes. In this case it is natural for the next shift to compensate for these production losses.

3.3. Organisational Learning

As has already been pointed out, it is necessary to look at the new department as an open system constantly interacting with its outer environment. An important part of that environment is the management system within the company. Production management has traditionally relied primarily on:

- MTM-based work standards
- prescribed work methods
- detailed work and control instructions
- minimum training limited to the performance of manual tasks

As mentioned earlier, this production philosophy has in part been questioned by the supervisor and the workers within the department.

- MTM-based work standards are still used but could only cover part of the work, since work in this case means more than just performing manual tasks
- Workers are encouraged to try new work methods
- It has been considered of great importance to maintain an ongoing training and learning process

It has also been considered important by production management to allow for a limited over-capacity of personnel, which means a revaluation of the traditional goal, i.e.to minimise the number of personnel. To some extent the over-capacity is created by the department itself, for instance by reduction of absenteeism. One reason for allowing this over-capacity certainly has to do with the changing relations of manpower and capital costs. The need for a deeper training of personnel is another reason.

It is also important to note that the supervisor in charge of the new unit has support for his views from colleagues in other departments within the new part of the plant.
Some of the supervisors actually make up a sort of informal group, and together they have put increased demands and pressure on other outside groupings of "systems" to make adaptations and reorientations in line with the general philosophy.

For years managers and supervisors have been sent to courses in psychology and management. The real outcomes of these courses have often been questioned. By taking an active part in a search and learning process, as has been the case within the new factory, it has been possible for managers and supervisors to try new behaviours, to get feedback from colleagues, and thus learn about their own role and how to manage the department.

When speaking with managers today, the development of new work forms is conceived as a necessary step to keep up with the same production goal.Thus the basic elements in humanisation of work are today more of an integrated part of management philosophy. It is also clear that developing new work forms does not mean lowering the demands put on the people working in production. On the contrary - if we take the new unit as an example - what really has happened is that the demands have been raised. That is, not the physical demands, e.g. heavy lifting, but the demands that people must use their judgement, learn more things, and grow in experience. That this really is so has been confirmed when newly hired young people have been asked how they experience their work situation within the company. Those working within the new department think that their work demands much more in terms of learning and understanding. They are more critical of the training being given than people in other departments - despite the fact that they have actually been given more training. At the same time there is a significant difference in the general attitude towards work situation in favour of the new department.

It should finally be mentioned that the industrial engineers have taken a very active part in the organisational learning process. They have looked critically upon traditional principles and have certainly - in view of new experiences - modified their own working methods and tools.

3.4. Perceptions of Job

When considering the workers' reactions and satisfaction it can be informative to start with a comparison of the overall satisfaction with the work situation.

On average the workers in the new department are more satisfied than the others, the means being 4.1 and 3.7 respectively on a five-point scale where high values mean high satisfaction. Other cues about the workers' satisfaction can be given by their intentions of leaving the job. Although all workers have about the same perception of their possibilities of finding a new job - 75-80% - only 37% in the new department would actually like to have another job as compared to 57% in the traditional departments. Similarly, 25% in the traditional departments intend to change their job within the next 12 months, but only 16% in the new department (Table 2).

Table 2. Questions concerning job change

	Traditional units		New unit	
	Maybe	Yes	Maybe (%)	Yes
Would you like to have a different job than the one you have now?	-	57	-	37
Do you see a real possibility of finding another job?	50	25	58	26
Do you actually intend to change your job during the next 12 months?	19	25	11	16

The general picture is thus in favour of the new department. It now remains to be seen what the specific conditions are that contribute to this higher satisfaction. In Figure 2 the workers' perception of the work content is shown.

The effect of the changed work organisation is clearly seen in these ratings. The ratings by the workers in the new block department are more positive in practically all aspects. The smallest differences concern responsibility which is rated high in all departments but still somewhat higher in the new block department. This probably reflects that all workers regard their work as demanding responsibility irrespective of its content.

The workers in all departments also consider themselves to be independent in their work in spite of considerably different degrees of freedom in their tasks. For example, the workers in the new department are much less dependent on specified work pace and specific set of routines. The workers are also less tied to their workplace and the work is less structured. A possible explanation of the similarities in the ratings can be found in their interpretation of independence specifically meaning independence of supervision. All jobs require in practice very little or any close supervision.

The differences are particularly strong in aspects related to personal growth and learning in work. The possibilities for self - decision on and development of work methods and for use of one's abilities and knowledge are seen as much better. Work is also seen as less monotonous and as more interesting. The general picture is thus highly congruent with the values behind the change in work organisation.

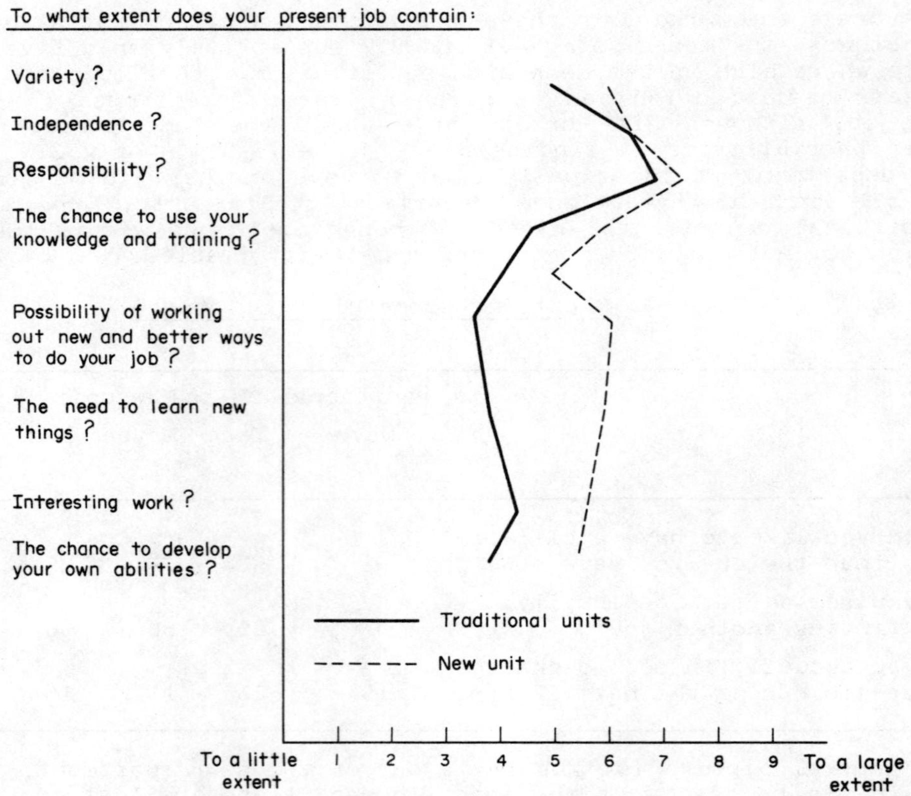

Figure 2. To what extent does your present job contain:

3.5. Social Relations

One important feature of the new work organisation is the need to co-operate and to solve problems in groups. This has been regarded as a favourable aspect of work, relations with co-workers being an important source of satisfaction. This group organisation is reflected in the workers' ratings: the possibilities to talk with each other during work as well as to help each other being better (Figs.3 and 4).

These improvments also depend on the lowered noise level and the lessened geographical confinement. More personal relations are, however, not affected. The personal interest taken in each other is the same in all departments.

The workers also rated the importance of the same aspects of work. The importance can be seen as partly reflecting the strength of the workers' needs.

All workers regard social relations as very important but the possibilities to talk to each other and help each other in work are considered slightly more important in the new unit.

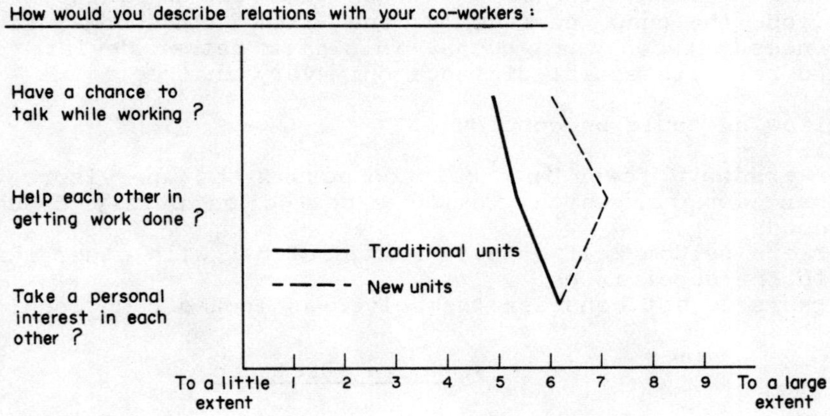

Figure 3. How would you describe relations with your co-workers:

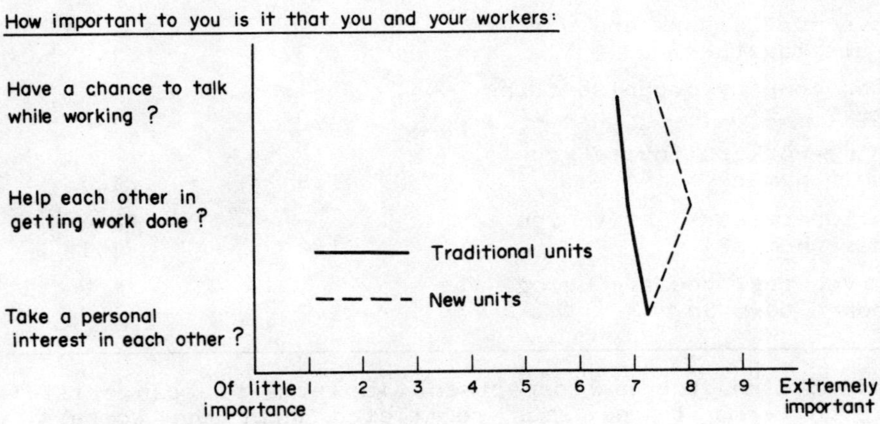

Figure 4. How important to you is it that you and your workers:

3.6. Social Distance

For questions related to feeling of social distance (see Table 3), the scale range is between 1-5. High values reflect more distance.

The social distance is small and the same in both the traditional units and the new one. Only one item (4) shows significant differences between the groups. This item rather deviates from the rest and reflects social distance only very indirectly.

The following could be concluded:

(a) Workers have fewer informal contacts with supervisors than with other workers, which can be expected on purely probabilistic grounds
(b) Workers seldom meet outside work, neither with other workers nor with the supervisors
(c) Workers do not consider themselves as looked down upon.

Table 3. Social Distance

	Traditional units	New unit
1. Do you have informal talks with co-workers?	2.6	2.6
2. ---with any member of the supervisory staff?	2.0	2.0
3. Have supervisors and workers meals together?	1.3	1.1
4. Does company organise trips, etc.?	1.9	1.5
5. Do co-workers invite you to their homes?	1.8	1.7
6. Do supervisors invite you to their homes?	1.2	1.1
7. Do you feel you are being looked down on?	1.2	1.2

The climate in the new department in terms of open criticism (see Table 4) seems to be less restricted when the workers and the foremen are concerned. The answers also reflect a strong correlation with hierarchical position; the possibilities being smaller the higher in position the target for criticism is. On higher levels the difficulty of access has the same effect on both types of units.

Table 4. Possibilities for open criticism

	Traditional units Yes (%)	New unit Yes (%)
Possibility to criticise openly:		
some of the workers	54	74
foreman	51	58
shop engineers	40	44
heads of departments	42	44
plant managers	32	33

3.7. Physical Working Conditions

Great efforts were made in the construction of the plant to minimise negative physical working conditions. The more convenient hours of work is a function of the present one-shift operation. According to the ratings such conditions have been much improved. Compared to other departments this new unit is less noisy, has better air conditioning, the localities being light and spacious. The work is also seen as less physically demanding. The psychological strain, on the other hand, is seen as about the same. The risk of accidents is also seen as slightly higher, probably depending upon new and complex equipment and a fairly large proportion of inexperienced workers.

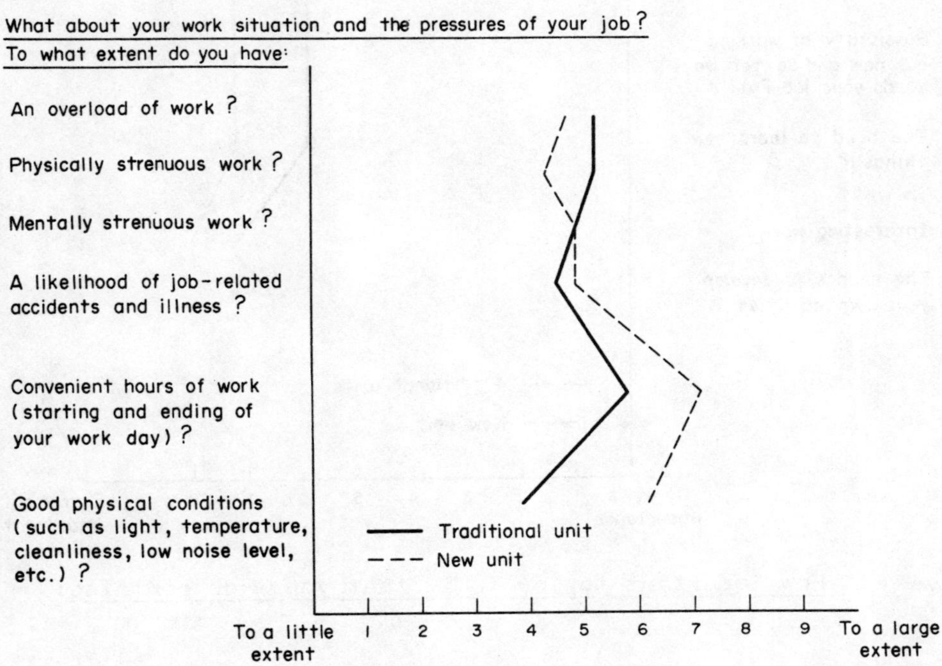

Figure 5. What about your work situation and the pressures of your job ?
To what extent do you have:

3.8. Self-actualisation

Ratings of the importance of the different aspects of work content should reflect the workers aspirations in these respect - their need for self-actualisation.

The workers of the new block department attribute more importance to all aspects of psychological work content. Indeed, these are important to all workers, but even more so to those who have experienced a different work situation. This would then indicate that the basic assumptions about human nature and needs behind these changes in work organisation are supported by the data. The need for growth is increasing when you get a chance to fulfil it. The aspirations of the workers are rising as a consequence of better opportunities. The rise in aspirations also shows in the ratings of employment conditions - Figure 7.

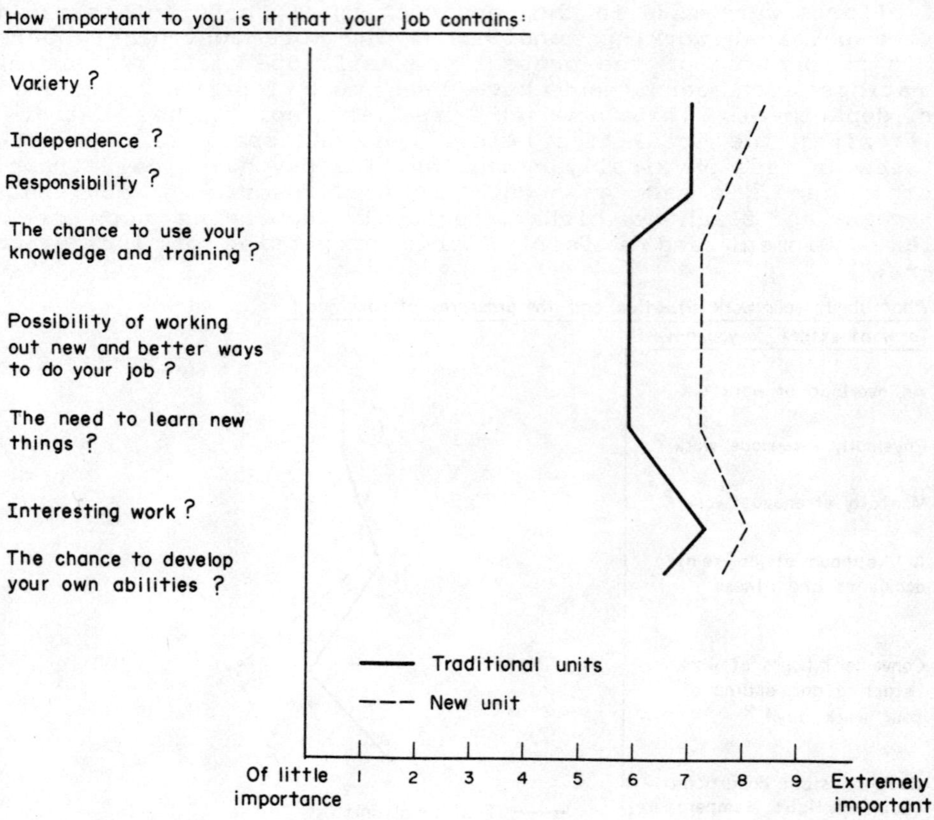

Figure 6. How important to you is it that your job contains:

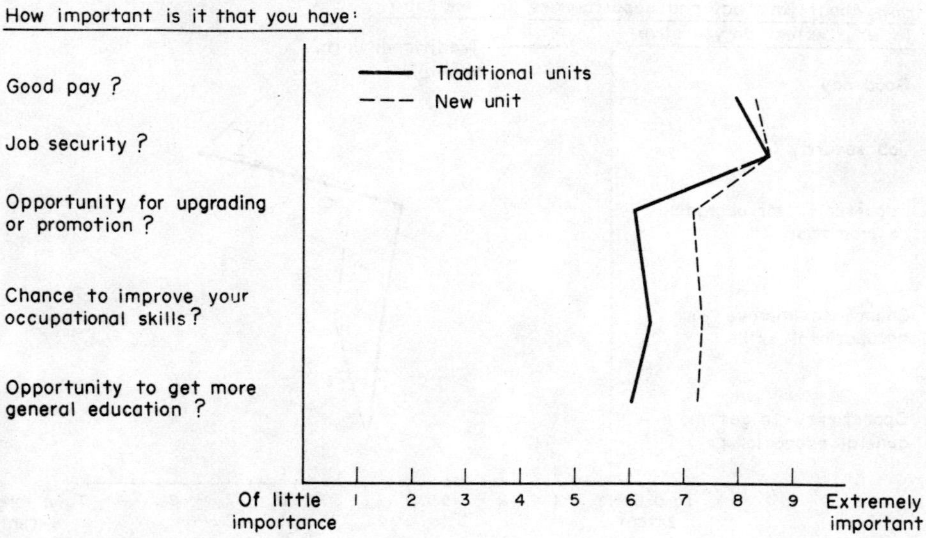

Figure 7. How important is it that you have:

The ratings are about the same in all departments for pay and job security - that is of equal and very high importance for all workers. But the growth-related aspects of employment are consistently higher for workers in the new department.

As has been reported here, strong emphasis has been attributed to the workers' opportunities to grow and develop in their work. The conditions for this are provided by the work organisation and learning in the daily work, as well as more formalised education in terms of shorter courses.

It is important to underline, though, that there are no formal educational requirements in addition to what is provided by the school system. Also the training required to be able to participate in work in the new unit is very short and does not deviate from that of other units - normally up to a month.

It is now reasonable to ask whether these generally more favourable ratings by the workers from the new department depend on a positive bias like a Hawthorne effect. Or is the social climate so good that the answers will be positive irrespective of what is being asked? It has already been shown, however, that there are deviations from such a general pattern, and this is even more evident in the following graphs.

In Figure 8 it can be seen that pay and job security are rated less favourably in the new department, probably due to the fact that the pay system had not yet been agreed upon and that at the time of the investigation some transfers had to be made. However, the conditions connected with personal development were rated higher, with one important exception. The opportunities for general education have not been affected by the change in work organisation. This shows clearly in the ratings which are not different between departments.

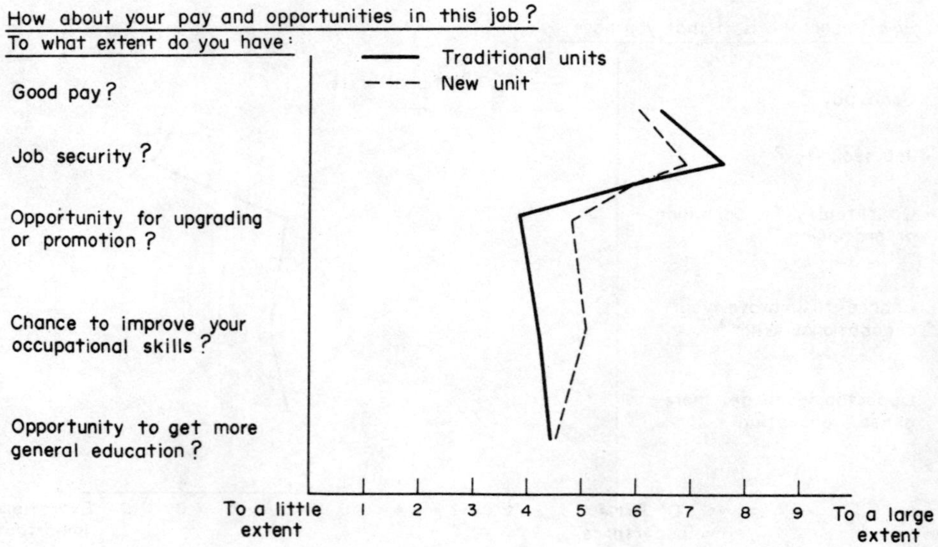

Figure 8. How about your pay and opportunities in this job ? To what extent do you have:

Figure 9 demonstrates a very clear case of consensus on certain values between departments.

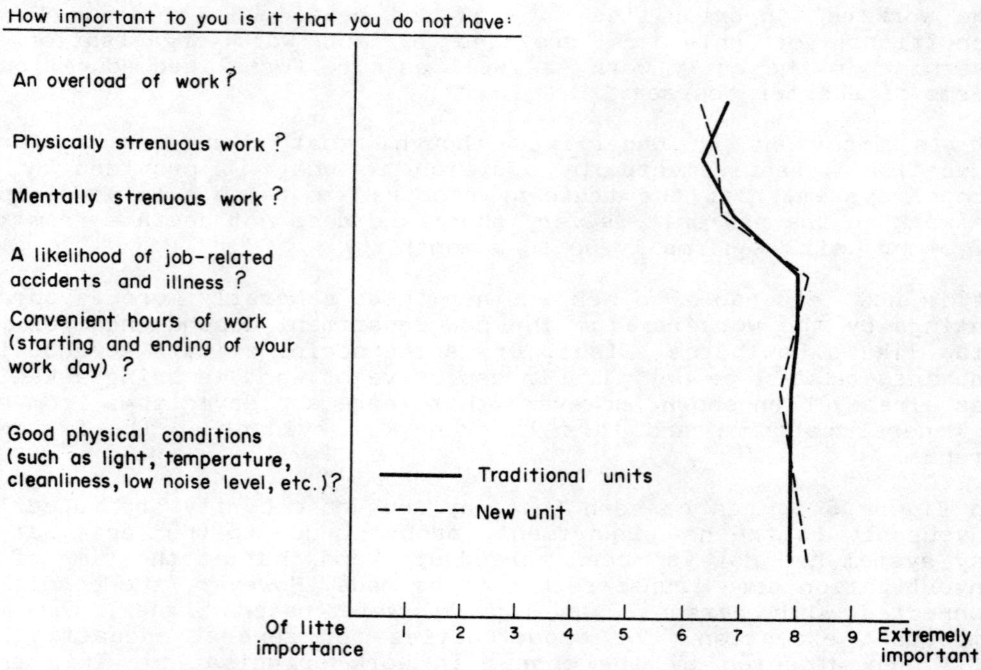

Figure 9. How important to you is it that you do not have:

In spite of very different working conditions the ratings from all departments coincide when the importance of these aspects are being rated.

This is further evidence of the validity of ratings, but it also reflects an important motivational difference between the psychological content of work and the physical working conditions. Herzberg would differentiate them as intrinsic and extrinsic factors. If the psychological content is improved, then the aspirations for personal growth increases (see Fig.6), but improvement of physical working conditions does not have a similar affect. Removing of physical nuisance in work is appreciated, but does not create a need for further improvements.

As three of the traditional units contain transfer lines - the technology under study - it is interesting to observe that the negative effect of this kind of automation was pointed out already in 1958 by William Faunce. With more automation and change in work organisation this effect can be reversed.

3.9. Attitude Towards the Company

As has been shown so far, the workers in the new department in most respects have a more positive attitude towards their work. A relevant question then is whether this positive attitude is generalised to encompass the company (Table 5).

Table 5. Attitude Towards Company [(a)]

	Traditional departments	New department
Is the equipment efficient and well maintained?	3.4	4.0
Does the company try to improve working conditions?	3.5	3.6
Is the work process sensibly organised at this company?	3.5	3.1
Is the management willing to consider the workers' ideas?	3.7	4.0
Does the company take an interest in the personal welfare of the workers	3.5	3.0

(a) High numbers indicate a more positive attitude on a five-point scale.

When the five-item scale is being accumulated, the traditional department and the new one show identical, rather positive attitudes. However, single items reflect minor deviations between the two groups. The fact that the department is new is reflected in the more positive valuation of the equipment. On the other hand, the efforts to improve working conditions do not show in the attitude towards the company. There is also a tendency to see the work process as less sensibly organised in the new department. The latter might be negative reactions from a few long timers rejecting the new less-rigid work organisation.

The effect of shorter tenure in the new department may show in the less-positive attitude towards the company's consideration of its employees. Thus it cannot be said that improvement of working conditions automatically gives a more positive attitude toward the company. Evidently a direct link between company policies and a changed work situation is not seen. But it is still an unsolved question why the improvement of working condition and work organisation is not reflected more strongly here.

3.10. Worker Participation and Attitudes Towards the Trade Union

In linking the social and the technical systems the result is to a high degree a function of the amount of influence from the social system. This influence can come from different sources: participation in the process of change by the individual worker, from demands from the workers' representatives via the union, from external specialists, or from the norms and values fixed by official institutions.

Table 6. Opinions about worker participation

	Traditional departments (%)	New department (%)
Workers should have the opportunity to participate in all matters affecting the company	42	58
Workers should have the opportunity to participate in matters affecting their own work	54	42
Opportunity to participate only when management asks	3	-
Not participate at all	-	-

From Table 6 it can be seen that in general all workers have high demands on opportunity for worker participation. For about half of the workers the demand for participation encompasses all matters while for the other half participation concerns only problems affecting the work. The claims in the new department are stronger.

Table 7. Willingness to participate in decision-making

	Traditional departments (%)	New department (%)
Willingness to take part in most decisions regarding the plant	21	26
Only on questions affecting my work	62	63
Only when management asks me	11	11
Not willing to participate	6	-

A part of this demand is evidently not connected with the worker's own participation. Table 7 shows that personal willingness to participate is less than the demand when questions above their onw work are taken into consideration. This might reflect that such issues are seen more as a task for the trade union. Willingness to participate can be said to be slightly higher in the new department.

Table 8. Personal participation during the last 2 years

	Traditional departments (%)	New department (%)
Participation in the meetings of the workers' organisations	21	16
Participation in different committees at the plant	15	11
Expressing ideas to your immediate supervisor	48	53
Expressing ideas to the workers' representatives	40	47
Contacting the workers' leaders outside the plant	6	5
Contacting governmental bodies	3	-
Arousing public attention	-	-
Methods of collective pressure	2	-

If we consider that workers think they should have opportunities to participate in decision making on all levels within the plant and that they are willing to participate on matters concerning their own work, have they tried to do something in order to influence managerial decisions? (Table 8).

In order to exert their influence, the workers mainly use two ways: expressing their ideas to the supervisor and to the workers' representative. Only 2% in the traditional department had used some method of collective pressure but nobody has tried to arouse public attention such as taking part in a demonstration or to strike.

It should be noted that about 95% of the workers are union members, but only 21% in the old departments and 16% in the new report that they are actively participating in the trade union meetings.

Table 9. Satisfaction with the union

	Traditional departments (%)	New department (%)
Fully satisfied	6	11
Somewhat satisfied	52	21
Partly satisfied - partly not	32	42
Somewhat dissatisfied	8	21
Very dissatisfied	3	5

Answers from the question regarding satisfaction with the union (Table 9) show some slight difference between the two groups, the trend being that workers in the new unit are more dissatisfied with the union. On the whole the answers reflect a rather unaffectionate attitude towards the union.

One issue of importance is to what extent and in what areas workers' influence should be channeled through their union representatives. Some hints on this question can be given from the items of importance of union activities and actual union efforts (Tables 10 and 11).

Areas that are closely related to workers' requirements of safety, i.e. wages, working conditions, and job security, are regarded as most important. These questions have also always been seen as the traditional union activities.

For these areas there is good correspondence between priority and perception of the union's real efforts. Improvement of working conditions form an important exception. The workers think that their union pays too little attention in this very important issue.

Regarded as less important are issues outside the sphere of working life. Here we can notice one difference between the departments. Seventy four per cent of the workers in the new unit think that the issue about general social and economic policy of the country is important for the union, whereas only 42% of the workers in the traditional ones see this issue as important.

Criticism is aimed entirely at the union doing too little rather than too much. Criticism that the union was doing "too much" could have been interpreted as showing that the personal influence of the workers themselves was too little.

How do the workers then perceive their possibilities to influence important matters, either by themselves or by the union? The answers on this question reflect both perception of their own possibilities as well as knowledge about the union activities. This means that such variables as tenure and involvement in the work of the union affect the answers.

Knowledge about the means of influence is reflected in the rather high frequency of "don't know" answers. On average one-third of all workers have given a "don't know" reply for each item.

Workers perceive the highest possibilities to participate in the following matters:

(1) Work organisation and working conditions (69%). The experience of participation in the change process is reflected in a higher percent of "yes" replies in the new department.

(2) Determination of basis and methods of wage payment (63%).

(3) Allocation of overtime work (63%).

(4) Workers' job classification and pay scale (61%).

Smallest possibilities regard promotion of workers (9%) and selection of management (4%) but also the planning of production and technical development - questions that are normally handled by staff functions.

This picture corresponds well to the traditional role of the trade union as primarily handling questions of pay and working conditions. It also reflects the worker's concern for questions affecting their immediate work situation.

3.11. <u>Interest in Social Affairs</u>

One important value in democratic societies is the involvement of the citizens in public issues. Again it is a question of what is "delegated" to the political institutions and what is of direct personal concern. The involvement is not only reflected in active political behaviour but also in emotional concern and intellectual interest. Societies have, to different degrees, managed to attract the interests of the citizens for central political questions. In many cases there is a strong concern about the alienation of the citizens in relation to societal matters, in others a more active population is feared.

Table 12 shows the average rating on ten central problem areas for which the workers have indicated their concern on a four-degree scale ranging from not interest to deeply interested.

Table 12. <u>How concerned are you about the following public issues?</u>

	Traditional department	New department
Problems of war and peace	2.4	2.6
Science and technology	2.3	2.4
Problems of economy	2.3	2.4
Problems of education	2.2	2.3
Problems of cultural life	1.8	2.0
Housing problems	2.3	2.1
Crime and delinquency	2.2	2.0
Health care	2.7	2.3
Pollution	2.8	3.0
Youth problems	2.6	2.4

The average over areas is 2.3, indicating a mean rating between rather and very interested. There is some variation between areas. Workers are most concerned about pollution, youth problems, war, and peace. Cultural life attracts least interest.

So far this is the picture of a passive interest with no practical obligations. The willingness for more active involvement is a function of the perceived possibilities to contribute to the solution. Almost 50% consider that they can contribute to at least some of the above-stated problems.

Among the means of influence, trade unions, resolutions from organisations, committee work, and contacts with public officials dominates. All of these are considered by 30% of the workers as possible ways of exerting influence. Strikes and demonstrations receive the lowest ratings - only about 10%.

In contrast to this rather optimistic picture stands the report of action taken. Only about 25% have tried more than once to exert influence on these matters. Also the frequency of active participation in voluntary organisations testifies to a rather passive commitment. Only sports associations attract any substantial number of followers - about 25%.

In conclusion then one can say that there is a big step between the rather high commitment to social problems on the theoretical level and the action taken. The important question here is what role does work play in the process of making workers active or passive members of society. The long period of very limited participation in decision making for workers in industry may have contributed to passivity. One can now speculate if a more democratic work situation with greater possibilities for direct influence also will raise the aspirations for active participation in politics on the community and society levels.

3.12. <u>Attitudes Towards Automation</u>

Technological change will always take place in industry and its consequences will be more or less beneficial for the workers. The attitude towards technical changes will to a large extent be a function of personal experiences from shifts in technology and how change has been brought about. In the new department the technology has to some extent been planned and designed in co-operation with the workers. Furthermore, change in work organisation has been connected with the technical innovations. It is, therefore, of interest to see if these differences in experience between units also have affected the workers' attitudes in this respect.

As expected, 100% of the workers in the new unit have experienced the introduction of new machinery. In the other units as many as about 60% have a similar experience. About the same number feel that they have been informed in advance - about 50%. Most commonly this information has been given by management. About 60% have been informed by managment and 40% by the union.

The effects of technical change on working conditions will be affected among other things by the workers' influence on choice of machines, organisation, etc. As many as 40% in all units did not know if there had been any possibility for workers' influence. In the traditional units 25% and in the new units 35% felt that there had been some influence from the workers. The rather low percentage of influence in the new unit reflects that few workers were employed in the unit during the construction period. These percentages definitely also show, though, that the workers' influence is very low in the important process of technical changes.

It can be seen that in general automation is expected to have little effect on questions concerning industrial democracy, whereas working conditions are thought to be strongly affected.

The general picture is in many respects very similar between the two types of units. The workers have a very optimistic view of the possibilities inherent in the technology. The idea that introducing new machinery will improve the conditions of the workers is held by 70% in the traditional department and almost 80% in the new. The support for the introduction of new machines is also strong. Fifty per cent of the workers in the new department would actively support technical changes and 30% in the traditional. Almost nobody would dislike or oppose such a change. This positive attitude is partly dependent upon the lack of fear for one's employment. Ninety per cent - irrespective of department - see no risk of losing their job because of automation, and the conviction is strong that they would be provided with a similar job in case of transfer.

There is a tendency for the workers in the new unit to be even more optimistic, particularly concerning job satisfaction, job safety, and physical strain. The conclusion is that workers who have experienced the more advanced technology along with new work organisation have a very positive attitude towards technological change. This is well illustrated from the observation that the workers in the new unit want further automation of the few remaining manual parts of the work process.

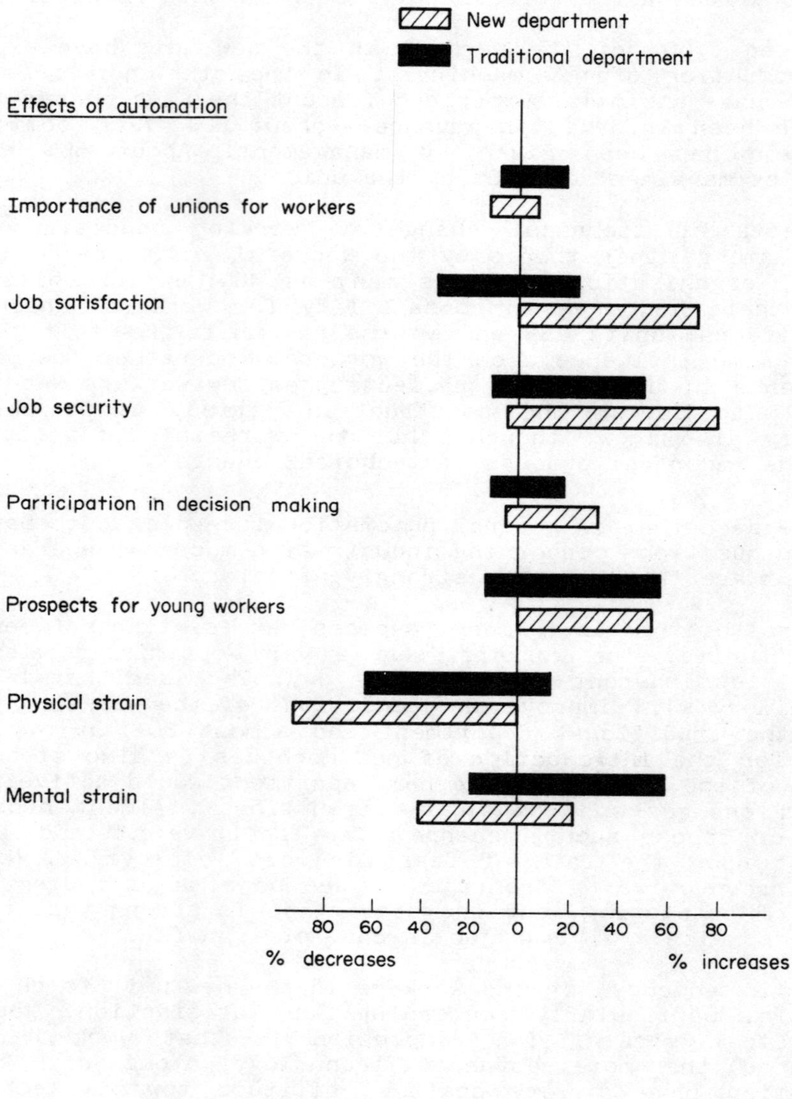

Figure 10. Effects of automation

4. CONCLUDING REMARKS

This report is an attempt to show that the social impact of automation in terms of qualities of the working conditions is not a function of technology *per se* but rather of the organisation used. The important message is that there are alternative ways of organising the work given certain technical equipment. The empirical findings support the view that a different work organisation based on other values and a different image of man is compatible with the demands of production at the same time as it provides more satisfying work.

The unit level has been deliberately chosen for the analysis rather than job level as being more in line with a "systems view". It is also based on the assumption that variations in jobs are caused by the work organisation. That is, different organisation gives different jobs and different content within these jobs. One organisation demands jobs like set-up men, operators, and maintenance men, all of whom have very different qualities. Another organisation does not demand these jobs but maybe only operators with enlarged responsibilities and task.So an analysis, in accordance with different job titles, will illustrate the effect of different job content. But job structure should be regarded as being as little given as organisation.

The concept of alternative organisation has important implications for the transfer of technologies between countries. The complex modern production requires that systems rather than single pieces of machinery are imported. These systems, as well as the machines, are often developed in a different cultural setting based on other implicit and explicit values. It is thus not just a value-free and neutral technology that is imported but along with it also a certain value system and assumptions about human nature. These values can be more or less in line with the values and conditions in the receiving nation. If there is a discrepancy normally it has fallen on the social system to adjust to the technology.

With a less deterministic view on technology, imported equipment and systems can be adjusted to fit with the social goals. They can then promote a process of social change in a direction that is often explicitly stated in the official policy of the country. The question then becomes the adjustment of automation rather than the social effect of automation. What, for example, are the processes that give the social system influence over the technical? And, finally, what are the roles of workers, unions, management, political institutions, and the producers of technology?

Table 10. How important should each of the following issues be in union activities?

	Very important Traditional department (%)	Very important New department (%)	Not important Traditional department (%)	Not important New department (%)
Improving wages, pensions, etc.	82	84	-	-
Developing cultural and educational activities in the plant	37	47	9	-
Providing recreational facilities	53	37	6	21
Improving productivity	25	42	10	11
Awakening a strong political attitude among the workers	26	32	38	37
Improving terms and conditions of work	80	100	1	-
Influencing the general social and economic policy of the country	42	74	13	5
Handling the workers' grievances	77	84	3	5
Ensuring the job security for the workers	88	84	2	-
Changing work organisation to increase job satisfaction	56	68	1	-

Table 11. How much attention does your union pay to the following issues?

	Too little		Adequate		Too much	
	Traditional department (%)	New department (%)	Traditional department (%)	New department (%)	Traditional department (%)	New department (%)
Improving wages and pensions	29	47	71	53	-	-
Developing cultural and educational activities	30	32	69	68	1	-
Providing recreational facilities	32	33	69	69	-	-
Improving productivity	9	17	89	83	2	-
Awakening a strong political attitude among the workers	32	21	65	79	3	-
Improving terms and conditions of work	51	68	47	32	2	-
Influencing the general social and economic policy in the country	32	32	66	68	2	-
Handling the workers' grievances	39	37	61	63	-	-
Ensuring job security	18	21	81	79	1	-
Changing work organisation	51	47	47	53	-	-

APPENDIX

1. WORK FORCE STRUCTURE

	Traditional units (%)	New units (%)
Age		
20 years or under	14.3	15.8
21-25 years	11.0	10.5
26-30 years	15.4	10.5
31-40 years	12.1	15.8
41-50 years	15.4	31.6
51-60 years	20.9	10.5
over 60 years	11.0	5.3
Sex		
Male	96.7	89.5
Female	3.3	10.5
Level of education		
Elementary school (7 years)	59	68
Comprehensive school (9 years)	31	5
County college	2	-
Grammar school	2	26
University	1	-
Correspondence courses and the like	4	-
Technical education		
Twenty persons in the traditional departments and nine in the new one had some kind of technical education		
Technical practical line of comprehensive school or 2-year technical continuation school	6	2
Technical college	2	-
University of technology	-	-
Others (correspondence courses and labour market training)	12	3

2. PAY STRUCTURE
Skr/hour 1975

	Traditional units (%)	New units (%)
SKr 17.24	4	-
SKr 17.24-17.50	-	5
SKr 17.50-17.94	14	26
SKr 17.94-18.75	32	11
SKr 18. 75-19.96	30	42
SKr 19.96-21.16	15	16
SKr 21.16-		

REFERENCES

Blauner, R. (1964) Alienation and Freedom, University of Chicago Press, Chicago.

Chinoy, E. (1955) Automobile Workers and the American Dream, Doubleday, New York.

Faunce, W. A. (1958) Automation in automobile industry. Some consequences for in-plant social structure, American Sociological Review, Vol. 23, p.p. 401-447.

Herbst, P. C. (1974) Socio-technical Design Strategies in Multi - disciplinary Research, Tavistock Publications, London.

Herzberg, F. (1966). World and Nature of Man, The World Publishing Company.

Kornhauser, A. (1965). Mental Health of the Industrial Worker, Wiley, New York.

Maslow, A. (1954) Motivation and Personality, Haper and Row, New York.

McGregor, D. (1960) The Human Side of the Enterprise, McGraw-Hill, New York.

Nobel, A. M. and Svensson, U. (1975) Automation och Industriarbete (Automation and Industrial Work), mimeograph.

Rice, A. K. (1953) Productivity and social organisation in an Indian waving shed; an examination of some aspects of the socio-technical system of an experimental automatic loom shed, Human Relations, vol.6, p.p. 297-329.

Seeman, M. (1959) On the meaning of alienation, American Sociological Review, vol. 24, p.p. 783-791.

Taylor, F. (1911) The Principles of Scientific Management, Harper and Row, New York.

Walker, C. R. and Guest, R. (1952) Man on the Assembly Line, Harvard University Press, Cambridge, Mass.

Wild, R. (1975) Work Organisation, Wiley, New York.

2. Automation and the American Automobile Worker Routes to Humanised Productivity

THE UNITED STATES RESEARCH TEAM*

The central problem addressed in this report is how automation of large scale industrial processes could be organized so that it would result not only in increased productivity but in a more satisfying and meaningful life for the worker.

The United States portion of this study identified indicators of work morale and productive behaviour - as well as of frustration and unproductive behaviour - that demonstrate the extent to which this problem is being confronted effectively in an industry vital to the social and economic health of the world's most productive country.

* The United States team which collaborated in this project consisted of Betty M. Jacob, Research Corporation of the University of Hawaii (project director); Philip E. Jacob and Arthur M. Whitehill, of the University of Hawaii, and George W. England, University of Minnesota (principal investigators); Richard C. Pratt, University of Hawaii (research associate); Chung-Si Ahn and Alex Farkash (research assistants).

This research was financially assisted by the US Department of State (Bureau of Educational and Cultural Affairs), the US Department of Commerce (Economic Development Administration), National Center for Productivity and Quality of Working Life, Ford Foundation, I R E X and the Johnson Foundation.

The United States team is also indebted to the United Automobile Workers, the General Motors Corporation, and the Ford Motor Company, which generously cooperated in the conduct of this study.

Specifically, the findings point out that:

- Most workers accept automation - but automation does not change their attitudes toward work or their performance on their jobs.
- The price paid workers to support productive technology is job security.
- Most of these workers are satisfied with their jobs - are reliable - and are productive.
- But the dissatisfied are <u>un</u>productive - often willfully absent, and ready to leave their jobs.
- Two profiles of work motivation emerge. One group exchanges time for pay, pleasant surroundings, and an easy work load. The other looks for self-expression through interesting work.
- Younger workers are the most "alienated" in the automobile industry. Their discontent challenges the most hallowed right of collective bargaining - the seniority system.
- A small core of socially conscious, union-oriented workers, who assume repsonsibility for both the work environment and civic activities, may be the change agents who can reconcile technological advance with improvement in the quality of working life.

Within these parameters of worker consciousness and disposition, management, trade unions, and those responsible for public policy in the United States seek routes to what may be termed "humanized productivity". Some of the options, in the light of the findings, are examined at the conclusion of this article.

1. THE DATA BASE

The main survey in the United States was conducted at two General Motors and one Ford Motor Company plants, using the common internationally constructed instruments designed to achieve cross-nationally equivalent measures of the key variables. The content of the job, organization of the work process, physical working conditions, extent and significance of worker participation, role of unions, attitudes of workers toward company and vice versa, and specifically the human consequences of automating production were simultaneously assessed by management, union representatives, outside investigators, and the workers themselves. In addition, basic aggregate data were compiled on the country's economic and social system, covering especially ownership, regulation and management of industry, labor legislation, social security provisions, and education.

The sample of workers was drawn from two departments in each plant - the automated transfer line in which the engine block was tooled by electronically controlled machinery, and the connecting rod floor where several similar machining processes were manually controlled. All major types of work performed in these production units were included in the sampling frame, including the main categories of operators, stock handlers, inspectors, "set-up" men, and maintenance personnel as well as the department supervisors.

All shifts were sampled. The 205 workers interviewed approximated the full population of workers assigned to the designated selection of work positions. Less than 5 % of those targeted were unwilling to be interviewed or were otherwise not available.

The limitations of this sample were (1) that it was drawn from a purposefully defined segment of the work force, i.e. employees in specified production departments in automobile engine plants, and (2) that the plants were not selected by procedures that assured they were a probability sample of all automobile engine plants in the country. On the other hand, within the specified universe, probability sampling procedures did assure that the workers who were interviewed effectively represented all major types of jobs in the designated automated and non-automated departments. This met a main requirement of the research design, that called for comparison of the effect of the "automatedness" of a job on workers' attitudes, perceptions, and behaviour, albeit within a single industry. (A parallel study in the steel industry, using similar sampling techniques and instruments, is nearing completion and will permit at least one check on whether findings are industry-specific.)

Selection of plants was necessarily governed in part by feasibility of access, being determined by arrangements that could be worked out by each company with local plant management, and correspondingly by national union leadership with the local union officials. However, the four plants chosen in the United States (including the one used for the pre-test) were among the major engine factories in the industry and were deliberately selected to allow for the possible influence of differences in size, methods of management, organization of the work process, location, and degree of modernization. In fact, there turned out to be little difference in the profile of responses that could be attributed to characteristics of the plant. We have concluded that the sample of workers from these plants probably did reflect the essential structure of this segment of the American work force, namely those engaged in the machining of engines and their components.

A demographic profile of the American sample (Table 1) shows that these men (only three respondents were women) were primarily a group of "steadies" - tied to their jobs by long continued service (averaging over nine years), and only 15 % of whom had experienced unemployment in the previous five years (not counting temporary lay-offs which they did not consider unemployment). Most were relatively well paid. (Only 4 % earned less than S 5.00 an hour base pay without adjustments for cost of living, overtime, etc. Twenty--five percent earned over S 6.20 an hour; half of the group were in the S 5.00-S 5.60 range). Almost half the group were above 40 years old; only 20 % were 30 or younger.

There was a wide spread in the group's education and technical training. About two-thirds had completed high school. But only about 20 % had gone on to college, and only 1 % went the course and got a degree. A quarter of the group had secured the equivalent of a year or more of technical training (usually in company institutes with company assistance). But more than half had no technical education or job training at all.

In comparison with the American work force as a whole, as sampled by the University of Michigan's Quality of Employment Survey (1973),[1] these producers of the automobile engine on the average were older, had worked their jobs longer, and were much better paid. A smaller number had gone to college. The proportion of blue collar workers was, of course, much higher than in the national sample, and there were virtually no women. These respondents belonged in what may be considered the male "elite" cadre of US industrial workers.

Table 1. Demography of the American sample

	Automation study $\underline{n}$ = 2o5[a)] (%)	Quality of employment survey $\underline{n}$ = 1496[a)] (%)
A G E		
Below 30	20	47
Below 30	80	53
S E X		
Male	99	62
Female	1	38
E D U C A T I O N		
Grade school or less	12	11
Some high school	20	14
Completed high school	49	38
Some college, college degree	19	36
P A Y		
Less than S 10,000	4[b)]	61
More than S 10,000	96[b)]	39
T E N U R E O N J O B		
Less than one year	10	32
One to five years	26	33
Over five years	64	35
C O L L A R C O L O R		
White	9	52
Blue	91	45
Farm	-	3

a) Over 95% of the targeted sample were interviewed in the Automation study; 75.5% of the targeted sample were interviewed in the Michigan survey.

b) Calculated on the basis of hourly rate of base pay of less than S 5.00 (not including cost of living and other adjustments or overtime).

2. THE RESPONSE TO AUTOMATION

A major hypothesis of our study was that the introduction of automated production processes would substantially affect the nature of the work (minimizing manual operations) and thereby alter the attitudes of workers toward their jobs, increasing their satisfaction and self-respect. Also, there might be a spill-over effect, resulting in greater civic responsibility and social concern of the worker.

Initial findings from the American data indicate that these assumptions must be sharply qualified. Workers attached to automated departments differ little in attitude and behaviour from those doing similar work in nonautomated departments. Thus operators tend to feel and act the same whether they work with automated or manually operated equipment. The same is true with maintenance and supervisory personnel.

What does make a difference is the specific content of the worker's own job. Maintenance men derive much more satisfaction from their work than operators, stock handlers or inspectors (see Tables 6 and 7). But it does not matter whether they service machinery in an automated or nonautomated production process. Viewed from this standpoint, one can conclude that the less automated his job, the more likely is the worker to be satisfied and productive.

There is one exception - the "set-up" position where automating the production of engines has created a new type of job with the worker determining the tools that are required, installing and changing them, and monitoring performance of the machinery. Here, but here alone, the original hypothesis seems to hold. With a change in the nature of work in terms of mental and physical requirements and amount of interaction on the job, "automatedness" has produced a more satisfied worker, at least as compared with the average operator.

Aside from this indirect and limited impact, however, the introduction of automation appears to have had little effect on work motivation. This applies also to workers' attitudes toward automation itself and their readiness to support more advanced technology. There was no significant difference in the percentage of those accepting automation either between departments as a whole or among respondents in the same job category in different departments. But a higher percent of maintenance personnel and supervisors were supporters than in the other job groups (93 % and 100 % respectively, in contrast to 70 % of the other respondents - Table 2), indicating that level of skill and responsibility has some effect on enthusiasm for advanced technology. Actually, most of the respondents, regardless of whether they themselves were working with automated machinery, accepted automation and were prepared to support the introduction of "work-saving new machines and equipment" into their plants - contrary to what many industrial relations experts anticipated only a few years ago. Only 17 % stated they would disapprove of such changes.

Table 2. The response to automation

Differentiating influences		Attitudes toward introducing work-saving new machines into plant		
		Active support	Qiet approval	Active resistance or quiet disapproval
		n=80	n=80	n=80
1. Type of job				
Operators	n=74	32	42	26
Set-up	n=32	28	41	31
Maintenance	n=42	43	50	7
Supervisors	n=19	95	5	0
Others	n= 3	39	52	9
2. Sense of job security				
Index A: fear of losing job because of automation				
The secure	n=166	48	39	13
The insecure	n= 31	6	52	13
Index B: expect company will provide other acceptable work				
The secure	n=135	47	39	13
The less secure	n= 60	27	45	28
3. Quality of physical working conditions: as perceived by worker, job provides good conditions, such as light, temperature, cleanliness, low noise level.				
to a large extent	n=48	46	46	8
to some extent	n=68	46	46	9
to a little extent	n=85	34	37	29
4. Job challenge: as perceived by worker, job provides variety, responsibility, independence, chance to use knowledge and training, chance to develop own abilities, etc.				
to a large extent	n= 69	51	41	8
to some extent	n=104	35	44	21
to a little extent	n= 28	39	36	25
5. Organizational climate: as perceived by worker, company tries to improve working conditions, takes an interest in personal welfare of workers, is willing to consider worker ideas,				

Table 3. Correlates of acceptability of technological change[a]

	Correlation of acceptability of automation
Workers' perception of:	
Job security	
- composite index of sense of job security (five items)	0.40
- afraid to lose job because of automation	0.30
- expect company will provide other acceptable work if displaced	0.20
Job challenge (variety, responsibility, independence, chance to use skills, learn new things, develop abilities)	0.30
Working conditions	
- physical (light, temperature, cleanliness, noise)	0.30
- safety	0.17
Organisational climate (company's concern for the workers' welfare, responsiveness to the workers' suggestions, maintenance of efficient machinery and work processes)	0.29
Communication	
- down	0.22
- up	0.30
Collegiality with higher-ups ("social distance")	0.40
Interaction with co-workers on job	0.20
Opportunity for promotion by merit (skill or hard work)	0.19
Social consciousness	
- the workers' concern for public issues	0.21
- his sense of political efficacy (ability to contribute to solution of public issues)	0.20
Overall satisfaction with job	0.23
Intended turnover (worker's intention to change job)	-0.18
Unemployment (as reported by worker)	-0.23
Pay (base rate as reported by worker)	0.23
Age	0.13

[a] For this analysis, a multi-item scale of acceptability of automation was used, combining the workers' evaluations of various effects of automation on workers with their response to the direct question asking whether they would support or disapprove of introducing work-saving new machines and equipment into their plants. The formula applied here is Pearson's product-moment correlation coefficient. Probability that any of these relationships would have occurred by chance is less than 6 in a 1000 (0.006) except for age where the probability is estimated at 0.03.

organizes work process sensibly, maintains equipment and resources efficiently.				
to a great extent	n=74	49	42	9
to some extent	n=91	38	44	18
to a little extent	n=34	32	35	33
6. Communication-down: as perceived by worker, he is informed beforehand about introduction of new machines and work processes				
Yes	n= 95	52	39	9
No	n=105	31	44	25

What appear to contribute most to this remarkable support for automated technology in the automobile industry are the following four factors. (For the statistical significance of the relationship of these and other factors to worker attitudes toward automation, see Table 3).

(1) A sense of security in the job is associated with support of automation. Of those who said they would strongly support introducing new, work-saving machinery into their plant, only 2 % felt they might lose their jobs because of automation. Eighty percent of the strong supporters of automation were confident that if their present job were taken over by the new machines, the company would provide them with "other acceptable work" (another 17 % thought the company might possibly do so). Put the other way around, 48 % of those who felt secure (who were not afraid of losing their job because of automation) strongly supported new machines compared with 13 % who would resist or "quietly disapprove"; of those who felt insecure (i.e. who were afraid), only 6 % were strong supporters against 42 % opposed. Of those who felt sure the company would provide acceptable alternative work if they were displaced by automation, 47 % stated that they would strongly support new machines, as against 27 % of those who were less sure (stating they thought the company might "possibly" provide alternative work, or definitely would not[2]. (Fig. 1)

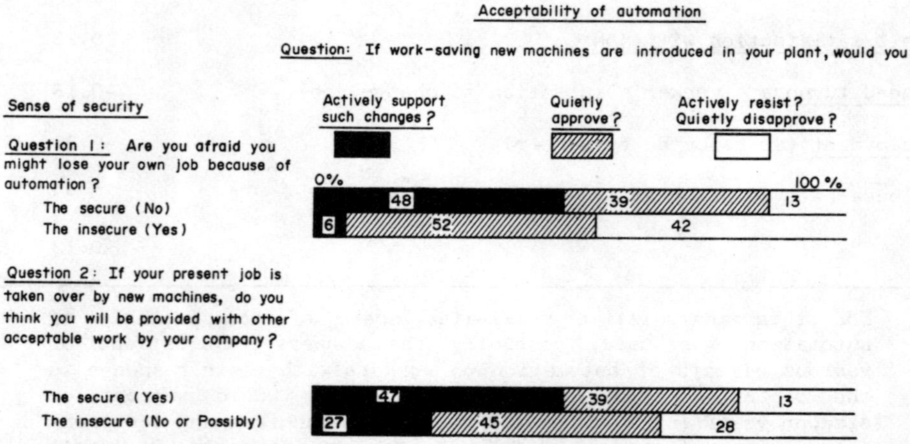

Figure 1. Sense of security and acceptability of automation.

(2) A second factor influencing acceptance of automation is the quality of working conditions. Workers who consider their working environment physically unsatisfactory are less likely to favor the introduction of new and more sophisticated machinery than those who consider conditions more satisfactory. Twenty-nine per cent of those who rate the physical conditions of their jobs as very poor disapprove of introducing new machines, as against 8 % of those who rate their physical conditions highly (Fig.2). In this connection, it is significant that one of the reasons most often given by the respondents in support of automation is that it will improve working conditions, especially in physical aspects - though workers in the poorer jobs evidently are not as sanguine about this prospect as those in the better ones.

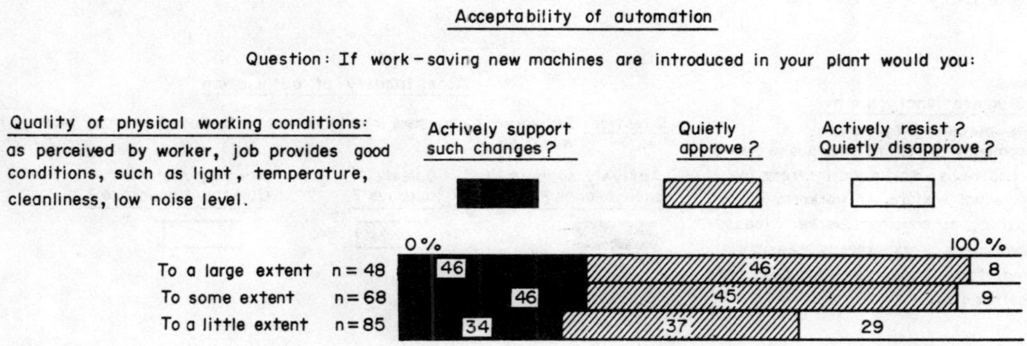

Figure 2. Quality of physical working conditions and acceptability of automation.

(3) Those that have the more challenging jobs tend to be more enthusiastic about automation. Paradoxically, these are the least automated jobs - skilled maintenance and supervisory positions involving less routine work and giving the worker greater autonomy in controlling what he does. (Consistently strong correlations - well above the 0.01 level of confidence - emerged in relating both single item and composite indices of the creative content of the job as perceived by the worker to the various measures of acceptability of automation and technological change (Fig. 3).(See Table 3)

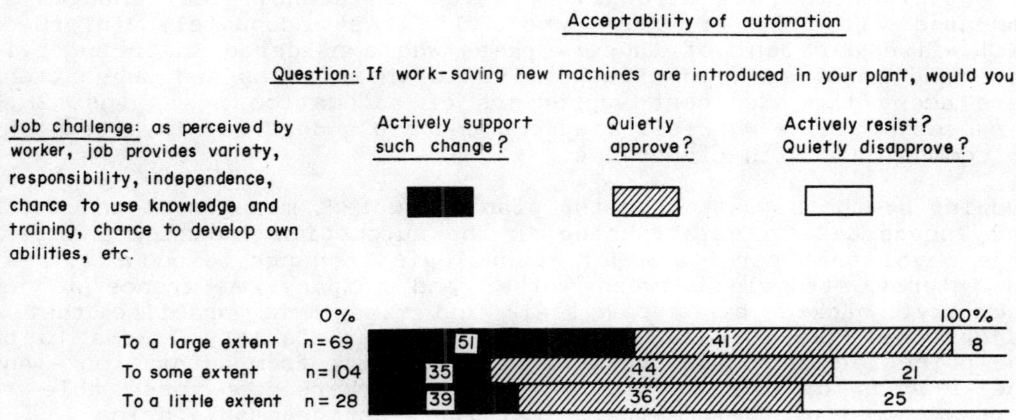

Figure 3. Job challenge and acceptability of automation.

(4) The management style, or "organizational climate," has an effect on a worker's attitudes toward automation. If he feels that management is concerned about his welfare, communicates with him and in turn is open to his suggestions, and runs an efficient shop, he will probably be more favorably disposed. A composite rating of company and plant based on selected items from Likert's instruments for evaluating organizational effectiveness was strongly associated with all of our indices of acceptability of automation (a correlation of 0.29 with the broadest index of acceptability, probability of less than 1 in a 1000 of this occurring by chance), see Table 3. Of those giving a high rating to the organization and management of their work, nearly 50 % responded strongly in favor of new technology; as against 32 % of those who gave a low rating to the company (Fig.4).

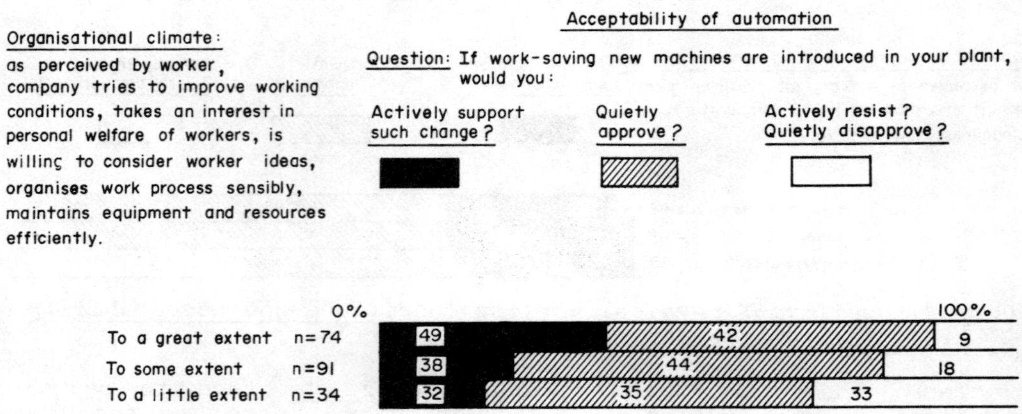

Figure 4. Organisational climate and acceptability of automation.

The most important component of this factor appeared to be openness of communication - up and down - between management and supervisors on the one hand, and the workers on the other. Fifty-two per cent of those who felt most adequately informed beforehand about changes affecting their jobs strongly approved of technological changes as compared with 31% of those who felt least adequately informed. Fifty-five per cent of those workers who considered their supervisors and management most responsive to their ideas and suggestions were among the strongest supporters of automation as against 26 % per cent of those who felt they encountered a deaf ear in attempting to communicate with higher ups.

Judging by the situation in the plants studied, management and union have succeeded in establishing in the automobile industry a tolerance level that permits major technological change to occur without an intense struggle between worker and company. Assurance of job security, backed by a pay scale and management practice that a majority of the workers find reasonably satisfactory, seems to be the price for increased productivity resulting from automation - and the leaders of the United Automobile Workers have been able to persuade most of their members that it is an acceptable price.

3. SOURCES OF WORKER SATISFACTION

The worker respondents in this study, who were still building the heart of the automobile at the time when the energy crisis had unnerved their industry, were as a whole satisfied with their jobs, reliable, and productive. These findings are generally similar to those of the comprehensive quality of employment surveys previously conducted by the University of Michigan's Survey Research Center with nationwide samples of the American work force[3] (Table 4).

Table 4. Satisfaction and productive behaviour

	US automobile engine plants sample	University of Michigan quality of employment survey:US national work forces sample	
	1974	1969	1973
	(%)	(%)	(%)
Very satisfied with job	37	46	52
Somewhat satisfied	50	39	38
Dissatisfied	13	15	10
Intend to change job (within a year)			
Yes, or very likely	10	16	16
Maybe, or somewhat likely	21	15	13
No, or not at all likely	69	69	71
Absent during last two weeks[a]			
- for any reason	13	-	19
- didn't feel like going to work	6	-	3

[a] This question was only asked in the United States

From the standpoint of productivity, the importance of these findings is the link definitely established between satisfaction and behavior. While physical output per manhour was largely standardized by the technology on the jobs studied, forms of unproductive behavior involving a cost element - such as turnover and willful absenteeism - were found to be directly related to expressed dissatisfaction of workers with their jobs. Overall job satisfaction was strongly correlated negatively, with intention to change jobs (r=-47).

The factor most significantly associated with job satisfaction was found to be the challenge of the job, though career possibilities, pay, job security, cordial relations within the work group, respect for supervision, perceived adequacy of physical conditions, and the organizational climate of the company were also important influences. The most satisfied workers were those who considered that their work had variety and was interesting, that it enabled them to develop their abilities and use their skills, and that they were called on to learn new things (Table 5).

Table 5. Sources of worker satisfaction

The workers' perception of:	Correlation[a] with statement of overall job satisfaction
Job challenge (variety, interest, independence, responsibility, mental demands, opportunity for developing abilities, using skills, getting more education, upgrading)[b]	0.62
Career possibilities[b]	0.52
Organisational climate (Likert measures of company's efficiency, concern for human resources, openness of communication)[b]	0.49
Supervision (adequacy of instructions, listening to worker, leaving worker alone, seeing he is justly rewarded)[b]	0.49
Pay (1) a worker's assessment of whether his job provided "good pay"	0.46
(2) actual gross base pay	0.25
Interaction with co-workers on the job[b]	0.42
Collegiality with higher-ups ("social distance")[b]	0.39
Adequacy of union[b]	0.38
Job security	0.35

[a] The formula applied here is Pearson's product-moment correlation coefficient. Probability of any of these relationships occurring by chance is less than 1 in a 1000 (0.001).

[b] These composite indices were constructed by first determining the inter-item homogeneity of relevant responses to the workers questionnaire; then computing the mean score of those items which correlated at or below the 0.01 level of significance.

An important component of job challenge is the extent to which it avoids monotonous and repetitive activities. As perceived by the workers themselves, monotony was strongly related to <u>dis</u>satisfaction (r=0,55); if the work was judged by <u>outside</u> observers to be highly repetitive, workers were likewise disposed to be dissatisfied (r=0.34).

A further component of job challenge was the extent to which the worker had control over the work process, i.e., "autonomy" in the sense of having independence and responsibility. The more control he felt he had, the more satisfied he was likely to be (r=0.41). This was confirmed by outside observations of the job structure: the more predetermined and routine the job program, the less satisfied the workers (r=0.30).

There was some discrepancy between the significance of physical conditions as perceived by the workers themselves and as assessed by outside observers. The latter showed a very weak relationship to whether workers were satisfied with their jobs, reflecting the fact that the outside observations were clearly not the same as those of the workers.

Given the high significance of work content (the challenge of the job) in generating worker satisfaction, it is not surprising that the degree of satisfaction differed greatly among types of job (Tables 6 and 7). The most satisfied groups were maintenance (those engaged in repairing the machinery), set-up (those responsible for installing and changing the tools), and supervisors. The most dissatisfied were the operators, product handlers, quality control inspectors, and similar workers, regardless of the type of machinery they were attending. The maintenance men more than any other group felt their jobs closely measured up to what they felt was most important in a job - except for physical working conditions. But then no one except supervisors had much of a good word for this aspect of their jobs, and even the supervisors were not happy with

Table 6. Satisfying and unsatisfying jobs

		Very satisfied (%)	Somewhat satisfied (%)	Dissatisfied (%)	Intend to change jobs (within year) (%)
Operators (both departments)	n = 74	19	62	19	16
Set-up workers (on the transfer line)	n = 33	46	45	9	3
Maintenance (both departments)	n = 42	64	34	2	0
Supervisors (both departments)	n = 19	47	42	11	5
Other (product handlers, inspectors, etc. in both departments)	n = 34	29	50	21	21

the physical conditions of work. Rather surprisingly, the supervisors, while they found their jobs challenging, were more dissatisfied than any group with their pay and felt almost as insecure in their jobs as did the operators.

Table 7. The dissatisfied worker[a]

Job facet	Operators	Set-up men	Maintenance	Supervisors	Others
Job Challenge					
Overall average of challenge facets	2.18	1.23	0.76	0.89	1.74
Variety	2.48	1.30	0.54	0.79	1.65
Interesting work	2.77	1.42	0.66	1.11	2.39
Chance to use knowledge and training	2.15	0.88	0.38	0.31	1.21
Possibility of working out better ways to do job	2.81	1.39	0.93	1.05	1.94
Need to learn new things	3.04	1.70	1.09	1.26	1.90
Chance to develop abilities	2.59	2.00	1.15	0.94	2.79
Independence	1.07	0.67	1.14	1.42	1.09
Responsibility	0.47	0.51	0.26	0.26	0.97
Job security	1.97	0.90	0.58	1.79	0.94
Good conditions	4.94	3.67	3.23	2.00	3.38
Pay	1.83	1.66	0.71	2.37	1.20

[a] Scores represent the mean difference between the importance worker attached to a particular facet of his work (on a scale of 1 to 9) and the extent to which he considered his job actually provided that feature (also on a scale from 1 to 9). The larger the discrepancy between desire and perceived reality, the more dissatisfied the worker was assumed to be; the lower the discrepancy score, the more satisfied the worker is considered to be. The original cross-tabulations from which these indices were calculated, were performed by Professor George England at the University of Minnesota.

4. WHY WORK? TWO PROFILES

Probing beneath the surface of the differences in degree to which workers were satisfied with their jobs, two types of person emerged, sparked by quite different incentives. One group evidently exchanged their time primarily for pay (and opportunities for increased pay), reasonable working conditions, and a reasonable work load. The other group's satisfaction stemmed more from opportunities for self-expression on their job, with interesting work the key variable.

This part of the analysis, undertaken by Professor George W. England of the University of Minnesota Industrial Relations Center, divided workers into two groups: (1) those for whom there was no significant positive relationship between what they saw to be the actual facets of their jobs and what they said was important for their job to provide, and (2) those for whom there was a significant positive relationship between what they felt their job actually provided and what they said was important for their job to provide. Then multiple regression analysis was used to determine which combination of 27 facets of the job best predicted overall satisfaction with the job for each group.

For the first group, six items combined as the best predictors of satisfaction. Four of the items hung together and shared variance: good pay, supervisor sees that you get justly rewarded, job provides opportunity for upgrading and promotion, and job provides a chance to develop your abilities. Two other items contributed unique variance: good working conditions and a reasonable work load. The results may be visualized as follows:

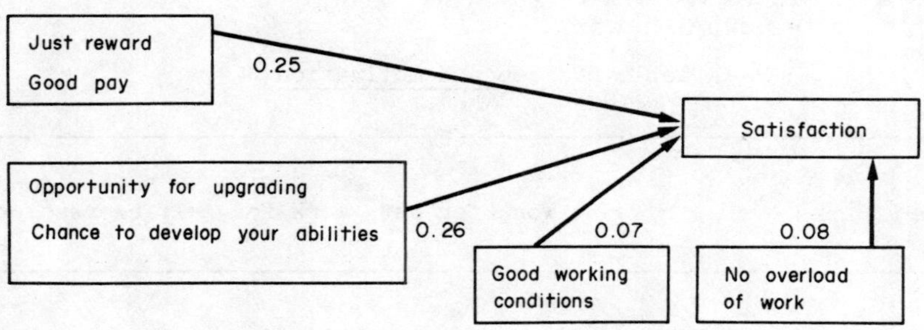

Profile 1. Work for pay group

Data for the other group of workers are quite different. Interesting work is the overwhelming predictor although it shares some variance with opportunities for upgrading and for workers to take a personal interest in each other.

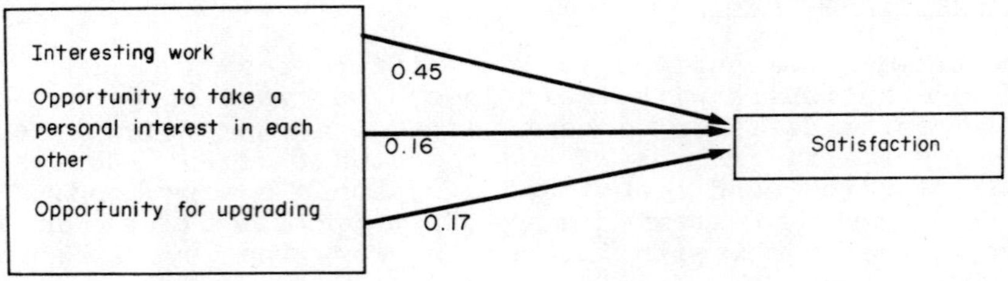

Profile 2. Work for self-expression group.

Both types of workers were found within every type of job, automated or nonautomated - at all age levels - and at all pay levels. However, as is evident from Table 8, more maintenance people appear to work for self-expression than for pay. The reverse is true for operators and other persons doing routine manual work like stock handling and inspecting. Supervisors and set-up men are almost equally divided.

This finding - that quite different incentives, in terms of work content and working conditions, operate to produce satisfaction in two groups - has major policy implications. Apparently no single, uniform, package of material rewards, job enrichment devices, and/or redesigned work processes will lead to satisfaction for everyone. For the first group, material rewards and conditions of work would be of primary importance in stimulating maximum satisfaction and effective work performance. For the second group, whose satisfaction is closely linked to psychological "self-actualization", increased opportunities for self-expression, largely through interesting work, are crucial factors. This is the group for whom participative management styles should work.

Table 8. Job and motivation

Categories	n	Work for pay	Work for self-expression
Total Sample	205	110	95
Department			
Automated	102	57	45
Nonautomated	85	42	43
Mixed	18	-	-
Type of Job			
Operators	76	45	31
Set-up men	33	16	17
Maintenance	43	17	26
Supervisors	19	11	8
Others	34	21	13

5. ALIENATION OF YOUNG WORKERS

On an overall basis, as we have seen, these automobile workers appear well satisfied with their jobs, intend to stay on with the company, and are ready to accept new production processes that increase productivity at the cost of jobs - though presumably not their own. (Fig. 5)

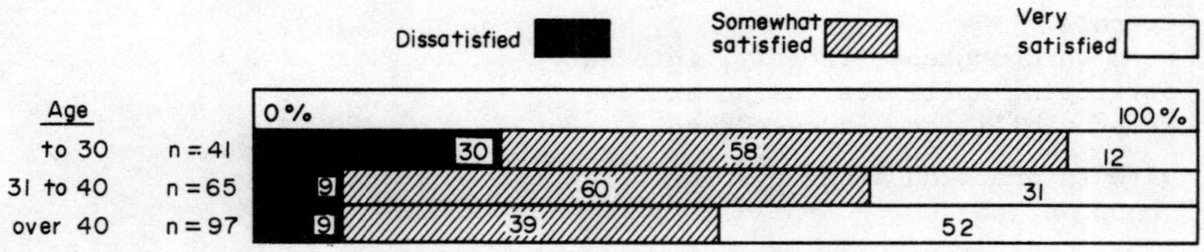

Figure 5. Age and overall satisfaction with the job.

But there is one group that is clearly and sharply out of sorts. These are young workers, aged 30 or less. Thirty per cent of this group expressed overall dissatisfaction with their jobs, compared with 9 % of older workers.

They are the least satisfied in every respect - with the kind of work they have to do, with their supervisors, with the company, with their relations, with fellow workers, and with the physical conditions in the plant (Table 9, items 1 and 2). They are in a word, alienated.

Table 9. Facets of alienation among young workers[a]

Correlation with	worker's youth
The younger the worker the more he expresses:	
1. Dissatisfaction with job	
overall	0.29
with work content (variety, interest, developing abilities, using skills, responsibilities, independence)	0.25
with interaction among his work group on job	0.33
with physical conditions (light, heat, cleanliness, noise,etc.)	0.27
with supervision	0.21
with career opportunities	0.19
with pay	0.22
2. Dissatisfaction with company	0.27
3. Dissatisfaction with union	
in general	0.34
in regard to specified activities	0.32
4. Sense of insecurity in job	0.33
5. Intention to change jobs	0.37
The younger the worker the less he:	
6. Expresses commitment to social goals	-0.25
7. Perceives positive effects of automation	-0.20

[a] All of these indices of dissatisfaction and other attitudes are associated with the age of the worker at a significance level of 0.003 or lower (estimated probability of association occurring by change = 3 in a 1000 or less).

What lies back of this? For one thing, younger workers tend to feel less secure in their jobs. Forty-two per cent of those aged 30 or less rated their jobs as providing little or at most moderate security as against 15 % of those over 40. Thirty per cent of the younger group stated outright they were afraid they might lose their jobs because of automation as against 12 % of the rest. Also, more had been recently unemployed: 24 % of the younger compared with 13 % of the others[4].

Another reason for their alienation is that the younger workers have the worst jobs - more monotonous, less opportunity to use their skills and develop their abilities. Sixty-eight percent were operators, 22 % performed other routine jobs such as stock handling or inspecting. None held a set-up position or was a supervisor. The remaining 10% were in maintenance. Forty-four per cent of the younger workers rated their jobs as very monotonous as against 13 % of those older than 30.

Adding to the aggravation of younger workers is the fact that they are over-educated in relation to the mental requirements of their jobs - much more so than the older. We calculated the difference between the amount of general education they acquired and the amount required for the job as estimated by foremen and standards engineers. This discrepancy was negatively correlated with age ($r=0.35$), meaning that the younger the worker the more his education exceeded the demands of the job. The size of this discrepancy in turn is a powerful variable in explaining most facets of job dissatisfaction and overall job dissatisfaction ($r=0.33$).

What is not a significant factor in making them dissatisfied, however, is pay. While younger workers tend to get paid less and to be less satisfied with their pay than the older (note the correlation of 0.22 in Table 9, item 1), the actual rate of pay appears to have no independent effect in determining the responses of different age groups. Within the same pay brackets, younger workers were still more alienated than older ones. In short, factors other than pay have to be responsible for these different attitudes of young and old. This was demonstrated in a series of cross-tabulations that related age to other variables while controlling for differences in actual pay.

All of this suggests that a major source of alienation among young workers may be the most hallowed of workers' rights in American collective bargaining - the seniority system - which puts them at the bottom of the totem pole. It is seniority that defers the rewards of choice jobs and - of greatest importance - job security. For the newest worker (usually the youngest) is the first to be laid off or furloughed when the market slumps for a new model or because of an energy crisis.

One might argue that this alienation is temporary. These young men are not permanently trapped in an inferior and underprivileged position. In due course they will move up the seniority ladder and enjoy all the privileges they are now denied. They are merely going the route their elders formerly travelled. But this younger generation appears impatient with time.

The consequences of such alienation are serious not only for the worker himself but for the company, for the unions, and for the well-being of the society. From the standpoint of productivity, of immediate significance is the strong inclination of the younger worker to leave his job and company. As seen in Table 10, only 24 % of those 30 or below stated they had neither the desire nor intention to change jobs, as compared with 59 % of those above 30. (In Table 9, item 5, correlation between youth and intended turnover is 0.37.) What makes this finding even more striking is that employment in each of the three plants in the inquiry was considered prestigious in the hierarchy of blue collar jobs in these communities.

The young workers tended to be less sanguine than their elders about the benefits of automation and hence about the welfare implications of technological advances. A significantly larger proportion said they thought automation generally did <u>not</u> increase job satisfaction, prospects for young workers or participation of workers in decision making (see Table 9, item 7). While this has not yet resulted in more outspoken opposition to further automatization, it does augur the classic antagonism of worker to machine where he sees in automation an enemy rather than the liberation of men from the dehumanization of mass production. After all, the role of these particular workers in the new technology appears as boring and personally meaningless as any assignment on the assembly line - and if the new machinery does increase productivity at the expense of jobs, it is these fellows who are likely to be out hunting for work.

Table 10. Age and turnover

Age		Strong desire and intention to change jobs (%)	Desire but uncertain intention (%)	No desire or intention to change (%)
to 30	n = 41	15	61	24
31-40	n = 65	11	46	43
41-50	n = 65	11	23	66
Over 50	n = 33	3	21	75

The unions are feeling repercussions. These younger disgruntled workers are less satisfied with what the union is doing than older members. More so than the older workers; they feel the union is doing less than it should in improving wages and working conditions, handling grievances, assuring job security, and changing work organization to increase job satisfaction (see Table 9, item 3). They tend to participate less in union activities; are less concerned with the union's taking a hand in shaping broad public policy; and do not think the union should spend much time stimulating workers' political consciousness. They themselves tend to be apathetic toward social goals and issues (see Table 9, item 6).

What seems to be happening is that the productivity pipeline is being sustained be the oldtimers, while the ingredients of explosion are being allowed to build up, almost unnoticed, as the young workers enter. The potential assets of this group - their energy and drive, their greater educational background, the long period of productive service that could lie ahead - are in danger of being lost in frustration and anger unless new patterns of work organization and other means can hold out rewards of challenging jobs at a much earlier time than now appears possible to these active, young recruits to the industrial ranks. They want an immediate entry into satisfying jobs. They want their "goodies" now, not after their turn has come on the slowly turning wheel of seniority.

6. LEADERSHIP THAT COUNTS

Constructive solutions to the problem of reconciling technological advance with improvement in the quality of working life may hinge upon a socially conscious, union-oriented core of workers who assume responsibility both for the work environment and the condition of civic life.

There is a significant overlap - at least among the sample of automobile workers whom we interviewed - between those who actively try to influence decisions affecting their work and those who are "activists" in the political and social processes of their communities, dealing with public issues. The link between the two worlds is the union, in which these activists are also principal participants (see Table 11)[5].

Table 11. The activists

Job activists[a]	Union activists[b]		Civic activists					
Workers who are engaged in:	Percent who are engaged in		Percent who are engaged in action on public issues[c]			Percent who take active part in organizations[d]		
	Three or more activities n=129	Less than three activities n=72	Frequently n=28	Few times n=119	Once or never n=42	5 or more n=25	2-4 n=81	1 or more n=95
Four or more activities n=73	88	12	27	64	9	25	33	42
Less than four activities n=128	50	50	9	61	30	5	37	58

a Workers who say they have personally tried to influence managerial decisions during the last two years by participating in the work of different committees at the plant, expressing ideas to supervisor or management, expressing ideas to local or higher level union leaders, participating in union meetings, contacting government officials, trying to arouse public attention, or engaging in methods of collective pressure.

b Respondents who say they have participated in union meetings, spoken at union meetings, talked to union representatives about public issues, or generally taken a regular and active part in union activities.

c Respondents who say they have attempted to contribute to a solution of public issues by talking to union representatives or party or government officials, writing to legislators or newspapers, speaking at union meetings, securing resolutions from organizations of which they are members, or participating in demonstrations.

d Workers who say they regularly take an active part, on a voluntary and unpaid basis, in political party, union, or youth, municipal, sports, education, religious, cultural, and other civic organizations.

A striking characteristic of this group is that they believe that what they are doing is of importance. Participation to them is prompted by a feeling of efficacy. They have not been turned into cynics by their participatory experience. These are no soured liberals. They are live models of the democratic "myth" - if you take on responsibility, you will indeed be able to make a difference. And they apply their belief in responsibility at one and the same time, as we have seen, in all compartments of their lives - at the plant, in the union, and out in the community.

Fortunately, not all of these activists are oldsters. A dent has been made in the corps of younger workers, some of whom have evidently decided to try the path of participatory expression rather than to succumb to alienation at least for the time being. The gamble, of course, is whether there will be a quick and visible enough pay-off to sustain the sense of efficacy and give credibility to the democratic conviction. Will their activism result in changing the conditions of work? Will union democracy make the rank and file feel that they have more than a nominal say in what happens? Will the public-spirited worker find responsiveness from the political and community establishments to his concern for bettering his society? Whatever the odds, this group of workers is at least willing to give it a try and put their own efforts on the line to accomplish it.

7. AREAS OF MANEUVERABILITY

The pay-off of this study is assessment of action that those responsible for the utilization of human resources in big industry can undertake to improve the quality of working life while increasing effectiveness in production. In other words, we are seeking the "area of maneuverability" where policy makers - in management, unions and government - can move to maximize the human benefits of automating industrial processes.

The preliminary US findings tend to support the thrust of Rensis Likert and his associates in identifying a style of management that enhances the morale of the work force, brings about a positive outlook towards technological advances, and makes for more productive behaviour, at least in the middle and longer run. This system exhibits a concern for the personal well being of the workers, opens channels of communication both vertically and horizontally, and gives workers opportunities to participate in decisions affecting their work life[6].

As we have reported, job satisfaction, productive behavior, and ready acceptance of automation among our sample of engine plant workers, were tied to such factors as the challenge of the job, job security, relations with fellow workers, the quality of supervision, and communication.

These are key elements in Likert's package of measures of "organizational climate" - and all within the competence of industrial decision makers to manipulate. Looking at the broadly positive attitudes and performance of the workers interviewed, one might conclude that in the American automobile industry these factors had been successfully manipulated by management - with strong promptings from the union.

On the other hand, in the ranks of younger workers and among that substantial group of operators and others doing routine and monotonous jobs we found a degree of disaffection that could seriously undermine the human base of the industry's stability and effectiveness. This is reflected in the rather modest rating of management on a selected set of Likert indices, scores of the whole sample ranging on the average from 3.2 to 3.4 in the plants studied (median = 3.28), with the minimum possible score being 1.00 and the highest 5.00. Obviously there is room for changes in climate factors which will lead to a more highly motivated and productive work force.

What can be done?

Five routes to what might be called "humanized productivity" could be tentatively suggested in the light of evidence from this study.

7.1. Automate More of the Production Process

This appears an acceptable route to nowhere insofar as altering work motivation is concerned. While most workers will not object if management chooses to introduce more sophisticated technology to increase productivity, there is no basis to expect that this will change the things that make work satisfying, at least in the kind of heavy machining operations involved in an industry such as manufacturing automobile engines. Here there is little room for maneuver in job design or work organization without total reconstruction of the plant in line with imaginative human engineering. More automation may add a few more set-up positions and pose new and interesting problems for the already highly challenged maintenance personnel. But routine remains routine for many of the tasks that still require the presence of a human, even if technology has succeeded in substituting electronic control panels for manual manipulation.

7.2. The Pay and Comfort Route

There is no doubt that adequate pay is imperative to strong work motivation. Recalling Professor England's analysis, it is evidently the prime factor for a particular type of person, along with aspects of "comfort" - pleasure, safe and healthy working conditions, convenient hours, and an unpressured work pace.

But for others these are a floor, not a ceiling, of aspiration. They are necessary, but not sufficient, conditions of satisfying work. And, given egalitarian pressures in a democratically organized union, pay differentials tend to be narrow and pegged to seniority. The contract, by majority choice, is geared to get adequate pay for everyone rather than high pay for those who work the hardest and are the brightest and the best. This is probably why income turns out to be at most a secondary explanation of differences in most of the attitudes and behaviors we were investigating. Pay lacks incentive force because there is little the worker himself can do to get more pay.

Most respondents as a matter of fact considered their pay rather good, with little complaint from either the most skilled or the least, the highest paid, or the lowest. Contributing to this state of modest euphoria was the position of these workers relative to others in their communities: their jobs were considered topdrawer.

As for the material conditions of work, it is safety and health - not comfort - that concerns most of the workers. Easy work is at the bottom of their shopping list of what is desirable in a job; and they are prepared to accept considerable physical discomfort if it seems an unavoidable part of a job that is otherwise satisfactory to them. But they are highly sensitive to hazards that remain uncorrected and to sheer disagreeable conditions that they feel could be improved - bad ventilation, filth, excessive heat, congestion.

Our data demonstrate a significant gap between workers' appraisal of their environment and that made by standards engineers and foremen. It is the former that counts as an important determinant of satisfaction and worker motivation[7]. The fact that there is no significant connection between the way workers and engineers perceive the situation suggests a possible area of direct maneuverability: here concrete, physical changes could narrow the perception gap and predictably raise the motivation level. On the other hand, a worker's general attitude toward his job, rather than the objective working conditions themselves, may have influenced the way he sees the conditions. In this case, a broader strategy of improving the overall quality of the working environment would be advisable.

7.3. Job Security

Here one comes close to the jugular. The sense of security as we have reported is a crucial factor in gaining the workers' support for the technological advances that generate greater productivity. It is also a vital determinant of general job satisfaction ($r=0.37$) and of workers' respect for and loyalty to their companies ($r=0.33$ with a composite index of satisfaction with company)[8]. Also, the more the worker feels secure, the more productive his behavior; for instance, the less likely he is to look for another job.[9]

This refutes, at least for these workers, the popular cliche that security has a molly-coddling effect, producing work-shirkers. In the American economic system, where employment is guaranteed neither by government (as in socialist countries) nor by company (as with large companies in Japan), job security is an option subject to decision by management negotiating with the union. Our data conclusively demonstrate this is one area where the best minds in management, unions, and government should concentrate their energies to find a formula which will assure workers of job security as a first step toward humanized productivity.

7.4. Make Jobs More Challenging

At first look, the findings from this study appear to justify the current emphasis on "job enrichment" and "job design". Job challenge accounts for an important segment of work satisfaction in this sample of engine plant personnel. It is particularly influential with those whose major motivation is to find in their work a means of self-expression, with opportunities to use and further develop their skills and to feel that they have counted for something.

However, given the fact that we are dealing here with heavy industry, and with the production end rather than office work, technology imposes severe limitations on what can be done to make any of the positions more challenging. Intrinsically, the maintenance jobs already meet the tests of variety, interest, problem-solving demands on skills and ingenuity, and independence. Also, the electrician, fixture repair main, or pipefitter is fully aware that every minute of shutdown costs thousands of dollars in the case of machinery on the transfer line - so on "trouble calls" he and everyone else knows that he counts.

But how do you enrich the work of a quality control inspector - without eliminating the job entirely through automatic electronical control? How does management redesign the job of an operator monitoring a Cincinnati broach or a bank of Excello borers, to add variety or introduce a creative learning process?

Our study did indeed find a clear and powerful connection between the content of the job and the enthusiasm of the worker for his job. But especially in <u>automated production</u> processes, the machinery to a large extent dictates the nature of the human work that is required and this pushes the search for motivation into other channels.

7.5. <u>Expand the Range of Worker Participation</u>

Many of us who helped to plan the multinational study expected that workers who had a greater share in determining what they did, how they did it, and what they got for their work would be more satisfied and more productive. We anticipated that participation in decision making would emerge as a key variable influencing worker motivation regardless of differences in the nature of their societies or the type of technology with which they worked.

The American data, however, hardly support this expectation insofar as these workers in a major mass production industry are concerned. We explored the <u>opportunities for participation</u> in decision making in the plant as perceived by the workers, and the amount and character of their <u>actual</u> participation. Neither turned out to be significantly related to most indices of job satisfaction, productive behaviour, or acceptance of new technology.

Opportunity for participation was assessed by asking the respondents whether it was possible for the workers of the plant to take part directly, or through their representatives, in decisions on 15 issues affecting their work: work organization and work conditions - job classifications, pay scale, and bonuses - hiring, lay off, and discharge - disciplinary measures - transfers to other work places - introduction of new machinery and equipment - cutback of production - and long range production and personnel planning including future work force demands.

Then the workers were asked which of eight activities they had personally participated in during the last two years (see Table 11, note a).

Table 12. Participation:The opportunity and the act

Amount of actual participation		Perceived opportunity to participate much %	some %	little %
five or more of the eight activities considered	n=28	42	36	21
four activities	n=45	27	33	40
three activities	n=56	14	45	41
two activities	n=51	22	33	45
none or only one activity	n=19	11	26	63

The amount of actual participation was related to the amount of opportunity workers perceived available in their particular plants (Table 12). It was also related directly to the workers' rating of the opportunity afforded in his own job to work out new and better ways of doing it (r=0.22); to the amount of responsibility his job entailed, as judged by outside observers (r=0.19); to his experience in having suggestions considered or accepted by management for changing the machinery or work process (r=0.21); to his satisfaction with the amount of interaction among co-workers (r=0.18); and to his feeling that it was possible for workers in his plant to openly criticize (in the presence of others) their foreman, shop engineers, or other management officials (r=0.18).

Participation also reflected an individual's own activist disposition. If he was active in union or civic affairs he was more likely, as previously reported, to be a participator in activities designed to directly influence decisions concerning his work (see Table 11). Also, if he had a sense of political or social "efficacy" - a feeling that his activities counted for something in trying to deal with social issues - there would be a better-than-even chance that he would take a hand where he could in decision making related to his job (r=0.25).

The puzzle of this situation is that participatory experience seems to produce so little visible feedback affecting either the dispositions or the performance of workers. Active participants are no more likely than passive time-clock punchers to find their work more satisfying or to hold a higher opinion of their company; on the other hand, neither are they notably more disgruntled or rebellious. There is no significant difference in their rate of absenteeism or in their intent to change jobs. As for attitude toward automation and technological change, the more active participators display slightly less enthusiasm than others.

Compounding the puzzle is a complete divorce of participation from any sense of the value of worker participation in general or of personal commitment by the individual worker. Respondents were asked whether they thought workers should have the opportunity to participate (1) in making decisions on all matters affecting the plant, (2) only on problems affecting their work, (3) only when management asked them, or (4) no opportunity at all. Then they were asked which of these alternatives would reflect the limit of their own willingness to participate in making decisions in their plant. There was virtually zero relationship between their responses to these questions and the degree of their own actual participation. On the other hand, the expressed desire for substantial participation in decision making and a feeling that workers should have the opportunity to do so seemed to have some roots in <u>dis</u>satisfaction, insecurity, and frustration: correlations of moderate significance were found between the verbal exponents of participation and those who expressed <u>least</u> satisfaction with their jobs (r=0.17), with company (r=0.19), with supervision (r=0.14), with pay (r=0.17), with safety (r=0.16), with work schedule (r=0.24), and with working conditions in general (r=0.17).

What this may suggest is that two countervailing influences are at work among this particular segment of workers which neutralize the motivational power of participatory experience. Those who have opted for participation within the framework of given opportunities are not propelled by any urgent sense of need. On the other hand, those who feel the need do not see in the <u>available</u> opportunities the means for effective action to fulfill their needs. For neither group does participation - <u>within the range of current practices</u> and in the scope of what these workers have experienced or imagined to be practicable - amount to a vital channel for the realization of their basic interests.

The range of participatory experience for these American workers may have been too small, and not enough concerned with the immediate circumstances of their own job and work situation. If they were not personally involved in <u>meaningful</u> participation (or knew others who had been) they might not imagine that it could provide substantial psychic satisfaction. There is some indication that this could be the case. The limited group of workers who had had the <u>most significant</u> participatory experiences (whether serving on joint management-union committees or mobilizing protest action <u>against</u> company policies) were more concerned than others about conditions of work and generally more satisfied with regard to their own jobs.

Another possibility is that collective bargaining is so well established in the United States that workers do not feel a close relationship between what they do <u>individually</u> and the conditions of work that satisfy. The system makes worker participation indirect, leaving him indifferent.

We await cross-national comparisons to figure out what may be happening here. Is the United States unique in this regard? Or is our experience replicated in some countries but not in others; and if so, what are the attributes of those situations where the participatory route to humanizing productivity is more promising?

Given the situation, what are the options in regard to the participatory route in big industry in America? One would be to forget it as being implausible and impractical. Another would be to offer participatory opportunities anyway - in the expectation that enthusiasm will grow with experience, especially if the participation is closely linked to the worker's own job. A third course would concentrate on the activist nucleus, who already demonstrate a high sense of social commitment, and build toward wider participation through an expansion of joint management-union collaboration, i.e., an American adaptation of the German model of co-determination.

What is clear is that participation is open territory for experimentation in the type of American plants included in this study. It is here that we can learn from the experiences of other countries, where fundamental reorganization of work processes seems, on the basis of case study reports, to have sparked worker initiative and introduced an entirely different outlook towards work. Bold and creative pioneering in human organization, as was accomplished in technological organization through automation, is a necessity of the time. To bring it off, however, will demand the shedding of concepts and practices deeply embedded in the structure of American industrial relations.

NOTES

1 See Robert P. Quinn and Linda J. Shepard, The 1972-73 Quality of Employment Survey, Ann Arbor (Michigan), Survey Research Centre, Institute of Social Research, 1974.

2 These findings conform to results obtained in a pre-test survey at an entirely different engine plant. For results of the pre-test, see Betty M. Jacob and Philip E. Jacob, Motivation for work and technological change, paper presented at the Ninth World Congress, International Political Science Association, Montreal, Canada, 1973. See also Report to the United States National Commission on Productivity, January 1973.

3 See Quinn and Shepard (1974), op. cit.

4 Correlational analysis shows a significant positive association between age and sense of job security meaning that the younger the worker the lower he rates the extent to which his job provides security (see Table 9, item 4).

5 The closeness of this three-way association between job activists, union activists, and civic activists is demonstrated by strong cross-correlations among the groups: job activism correlates with union activism at 0.52 and with civic at 0.40; union activism correlated with civic at 0.48. These are all well below the probability level of 1 in a 1000.

6 See Rensis Likert, The Human Organization, McGraw-Hill, New York, 1967; also his Human resources accounting, Personnel, May 1973. Instruments developed to measure the effectiveness of management systems by these criteria are described in James C. Taylor and David Bowers, The Survey of Organizations, Ann Arbor (Michigan), Institute for Social Research, 1972. A five-nation comparative study of the effects of differing systems of industrial organization, by Arnold S. Tannenbaum and others, has recently been completed: Hierarchy in Organization: An International Comparison, Jossey-Bass, San Francisco, 1974. A similar eight-country study is being coordinated by the European Coordination Centre for Research and Documentation in the Social Sciences (Vienna, Austria).

7 The workers' perception of the adequacy of physical working conditions correlates with general job satisfaction at 0.35 (significance = 0.001). Observers' reports (based in part on the standards engineers' evaluations) do not correlate at a satisfactory level of significance with worker satisfaction. Nor is there a statistically significant association between the workers' perceptions and the observers' reports.

8 These correlations are significant at the 0.001 level of confidence (probabililty of 1 in a 1000 or less that they could have occurred by chance). The index reported on here is the workers' general rating of the extent to which his job provided security.

9 There is a highly significant negative correlation between the worker's rating of the extent to which his job provides security and his declared intention to change jobs within a year (r=0.40).

3. Automation, Skill Structure and Trade Unions*

YVETTE LUCAS, CHANTAL BENAYOUN,
MONIQUE MEMBRADO, EMMANUELLE SOULA,
GILBERT FAYDIT DE TERSSAC

* The French part of the "Automation and Industrial Workers" project was initiated by Prof. J. Stoetzel and Mr. A. d'Auriac. Prof. W. Grossin (University of Nancy) and Dr. Y. Lucas (French National Scientific Research Centre, Toulouse) became the scientists responsible for the study in France. The A.F.E.E.T. (French Association for Studies in Work Environment) was created to support the research. Its president is Mr. Y. Delamotte. The C. G. T. expressed its interest in the project but was unable to raise funds. The C. G. T. - F. O. is involved in the A. F. E. E. T. but did not take part in the project. The project was financed by a grant from the French Ministry of Work, Employment and Population. Officials of this ministry have taken part in the research.

Note:
C.F.D.T.: Confederation Francaise Democratique du Travail
C.G.T.: Confederation Generale du Travail
C.G.T.-F.O.: Confederation generale du Travail Force Ouvriere

INTRODUCTION

The French study compared four units in the same plant. From this perspective it can only be regarded as a case study. Nevertheless, many of the results can be considered representative of conditions generally prevailing in automated and nonautomated production units in the automotive industry, even though they do not reflect variations in contextual plant conditions. Our present observations should be examined alongside other French studies on the same topic.

Our report will focus on the workers' attitudes towards automation, work content, working conditions, and work environment. However, the first section of our report will elucidate some general economic factors, characteristics of relations between employees and employers and the organisation of French trade unions which are necessary for understanding our results. For example, important sectors of the French market economy are nationalised, especially in the automotive industry; the trade unions, despite a small number of adherents, have great influence in the representative elections; and a great part of French labour legislation is a product of the class struggle of workers and trade unions. Another characteristic feature of our country is that, in general, the rise of automation was not regulated by State measures, and no committees or institutions were established to supervise installation of new techniques.

1. THE SOCIO-ECONOMIC CONTEXT OF AUTOMATION IN FRANCE

1.1. The State's role in industry

France has a market economy characterised by private ownership of most means of production. However, there is a State sector and a National sector of production. The intervention of the State in economic life manifests itself in financial aid to private firms, also in the development of the five year "National Plan". This plan is only programmatic; economic activity is determined by centres of decisionmaking. State intervention in the management of production has substantially increased during the past 15 years. This close inter-relationship between governmental bodies and private national as well as multi-national firms is a major feature of the French political and socio-economic system.

Some firms are fully owned and managed by the state. Nationalised firms are particularly important in the automotive industry: Renault, the leading producer of cars and trucks, is a nationalised firm. On the other hand, Peugeot, where this study was conducted, is a private enterprise and constitutes the second giant in the French automotive industry. Recently, some nationalised companies signed agreements with private companies. Such agreements exist between Renault and Peugeot in the automotive industry. This arrangement is a specific illustration of the current trend in the French economy. We may also note the participation of the state, or organisations depending on the state, in financing some companies. Help is also given by the State in the form of subsidies to private companies, e.g. to the International Company of Computers and Peugeot/Citroen grouping in the automotive industry as well as support to pioneer industries.

During the mid-50's the French economy had to face the problems of the emergence of the European Common Market. The competition with

West European countries, especially with the Federal Republic of Germany and Italy was very acute. At this time there was in France a period of change and rebuilding of many industries. Many firms, especially in textiles, had to buy new machines so as to increase their productivity rapidly, as well as their intensity of work. Others had to change their products or their manufacturing methods. This provoked a real growth of the economy and gradually technological unemployment was reduced. At the beginning of the 60's, the French economy was integrated into the Common Market and was stabilised, even growing towards the beginning of the 70's. But the upsetting of French industry in the 50's, although relatively far back, left an impression on workers' minds and led to bad feelings towards changes in technology with risks of job insecurity. This still appears in the workers' attitudes as the results of the study will demonstrate.

Since then, France has confronted one general crisis with various phenomena: inflation has grown very rapidly, there are some problems with the trade balance, unemployment has grown substantially. Now about one million people are unemployed, part-time work has increased and, at this writing, it was considered that the economic growth for 1975/1976 might be negative. The French automotive industry was at first strongly affected by the crisis. Now, the situation has improved especially due to the possibilities developed by increased exports. Concerning the firm studied (Peugeot), the problem was not so related to car production volume, which remained rather high, as to the oil crisis which provoked several changes in the character of the products. Thus, the long existing sector for production of diesel-engine cars in Peugeot has greatly expanded as a result of the oil crisis. This development had just appeared as we undertook our study and changed the relative importance of the different sectors of the plant.

As can be seen from the following tables, the output of the automotive industry increased up to four times since 1955, whereas the number of employed less than doubled. This increase in per-man output shows the economic impact of technological development in this industry which took place especially in the early 60's.

Table 1. <u>Rate of growth in the automotive industry expressed in units produced</u>*

Years	1955	1960	1965	1970	1971
Passenger cars	100	222	260	448	487
Trucks	100	144	129	156	170

* As a percentage of 1955

Table 2. <u>Percentage of the total economically active population employed in the automotive industry</u>

Year	1955	1962	1968	1970	1971	1972
Percentage	1.25	1.62	1.73	1.97	1.98	2

Despite the emerging crisis, the general situation of Peugeot production was good at the time of our study. The production plan was optimistic and the workers worked full-time. There was no special fear concerning lay-offs except among the foreign workers, whose possible dismissal could be regarded more as a rumor than a real danger.

1.2 French society and technological change

Since we consider automation as the independent variable in our study, we must take into account the attitudes of French society towards automation and technological changes, and to consider their relationship to the French socio-economic system.

Generally, adoption of automation came from decisions of management in French firms, especially to increase competitiveness and initiate new kinds of production. However, state subsidies to industry were not customarily provided for the purpose of introducing automation but as a result of other considerations such as offsetting the business cycle, transfer of plants to new towns or the merging of firms. No special institutions were created to promote the development of automation, though various governmental and private institutions and committees were involved in planning and guiding the introduction of automation.[1] There was no legislation specifically devoted to this problem.

Neither did trade unions consider automation as an issue per se, and, in general, they did not initiate actions directly related to automation. This may partially explain why automation does not appear as a distinct issue in the attitudes of the workers we have studied. French workers more often evaluate automation in reference to specific aspects of machine organisation, or limited effects on job content than in terms of its broader implications. However, in recent years, the C. G. T. has become interested in this issue considering it as a way to reduce the work load and physical strain and to introduce more modern means of production. At the same time, the C. G. T. regards capitalism as unable to solve the problem, and believes that a new way of combining intervention by the workers with a change in governmental policies is required, such as proposed in the Common Programme of Government of the left parties.

1.3 Trade unions in France

The trade union structure in France is a basic intervening variable affecting the social consequences of automation. It is characterised by three principal features:

(1) the plurality of trade unions at the national level and because of this also at the district, local, and plant levels. This is not a reflection of "corporatist" diversity but of different philosophical and ideological orientations;

(2) the small proportion of workers belonging to the unions, in contrast to their large importance in the workers' elected boards at plant and national level and in formulating national policy;

(3) the independence of unions with respect to political parties. Although parties and trade unions may have joined together to

press various collective claims, no trade union in France is direct affiliated with a political party. Many trade unionists in France do not belong to a political party; also members of the same union may belong to different political parties. However, recently the C.G.T. and C.F.D.T. have supported the parties of the opposition by following the Common Programme of the Government of the Left.

In France there are five big national trade union confederations which have been officially recognised: the C.G.T., the C.F.D.T., the C.G.T-F.O., the C.F.T.C., and the C.G.C. Originally there were only two, C.G.T. and C.F.T.C. (The Christian Union). Each split. The C.G.T. in 1947 divided into the C.G.T. and the C.G.T.-F.O., each belonging to different world confederations. In the 60's the majority of the members of the C.F.T.C. created the C.F.D.T. which does not consider itself a Christian Union. The minority remained attached to the old name (C.F.T.C.).

The C.G.C. (General Confederation of Cadres) was created more recently to organise permanent officials and managerial staff.

On the national scale, we should note also the existence of the F.E.N. (Federation of National Education), which is the most important union for teachers and officials of the national education system.

The workers' confederations (C.G.T., C.F.D.T., C.G.T.-F.O., C.F.T.C.) form <u>federations</u> for different branches of industry. There is no union specifically for the automotive industry in these confederations; automobile workers belong to the various Federations of Metal Workers, within which there are "motorcars" sectors.

Each local trade union in a firm, e.g. Peugeot, has a double affiliation: territorial (linked to a regional level union) and to the federation.

Table 3. <u>Union membership</u> (national level)

	C.G.T.	C.G.T.-F.O.	C.F.D.T.
1946	5.000.000		(C.F.T.C.) 390.000
1948	4.400.000	350.000	(C.F.T.C.) 430.000
1958	1.600.000	400.000	-
1968	1.900.000	500.000	C.F.D.T. 627.000
1971	2.100.00	550.000	678.000

Overall union membership (see Table 3 above) represents a very low percentage of the national labour force, but the percentage in particular socio-professional categories, branches of industry and localities varies greatly.

The trade unions designate representatives to the official commissions responsible for preparing the National Five-Year Plan, but their role is only consultative. The Social and Economic Council also includes representatives of unions, as do many other governmental commissions and boards. The existence of many of them is linked to the business cycle.

The role of unions is also very important in informing workers on social issues: through their role in the workers' elected boards in the firms; as negotiator on department, company, regional and national levels; by their adversary role in industrial conflicts which are very significant in the occupational life in France.

2. THE SPECIFIC CONTEXT OF THE INVESTIGATION

2.1. The plant

Peugeot Automobiles and the Montbeliard-Sochaux collectivity, where the French study was undertaken, is a private family business, expanding with the growth of the automobile. At present, the Sochaux Peugeot plant includes 36.000 workers. Car production in this plant is 41.000 monthly. The daily production is 2.253 cars. It was for a long time the sole industry in this isolated part of France and was, therefore, the supporter of numerous generations of workers, a situation that has still not entirely disappeared.

A French newspaper gives an impressionistic picture of the local atmosphere: "It is really a unique small town, where it is cold and rainy. And the rain slips down the stones and the concrete that covers the blocks. Heavy shapes and monotony reflect Peugeot's domination of the district. Montbeliard is a strange town. There, you work at Peugeot; you buy food in Peugeot's supermarkets (the RAVI); if you own a car it's often Peugeot. In short, you live on Peugeot's time and you can really never forget it. So 30.000 people bitterly realise that the money they earn at Peugeot they give back in one way or another to those who are paying them".[2] Also, many people rent their homes from Peugeot but are forced to leave if they lose their jobs or retire.

The collective agreement regulating plant life is the "Regional Agreement of Metal Workers". The last agreement was signed in 1955.

Some characteristics of the Peugeot plant:

(1) A management system which is considered to be paternalistic.

(2) The system of "Points Peugeot" in the case of transfer from one job to another, which is unique to this company. People get compensation if the new work pays less. They get a number of "points" which they may use whenever they choose. The C.G.T. and C.F.D.T. consider this to be insufficient and demand a reform of this system.

(3) The "Voluntary Saturday" which consists of supplementary working hours on Saturdays. In principle, people work voluntarily during these hours. This system was instituted by a contract which the C.G.C. and C.G.T-F.O. signed, while neither C.G.T. nor C.F.D.T. did so. The C.G.T. considers, as do many workers, that it is neither correctly paid nor truly voluntary.

(4) The suggestion system which has existed for a long time in the plant and is accepted by many workers, although trade unions such as the C.G.T. and C.F.D.T. generally oppose it.

2.2 The units

Four units were studied: two automated and two nonautomated. In the report these are designated respectively as A and NA units.

Table 4. Description of the units

Automated	Nonautomated
Unit 1: Production transfer line for engine blocks (6 to 10 years old)	Unit 2: Production conventional production of engine blocks
Unit 3: Production transfer line for engine blocks (more than 10 years old)	Unit 4: Tool-making shop

Unit 1 includes three transfer lines and some manual machines (43 operations) with 11 operators.

Unit 2 is composed of both automatic and nonautomatic machines performing conventional boring, drilling, milling, and grinding operations. Each machine requires one operator.

Unit 3 is composed of transfer machines. Some manual tasks remain such as manual assembling, disassembling, and pneumatic cleaning. Twenty three operators work at seven transfer machines including 134 operations, and at 16 manual workplaces including tool-machines.

Unit 4 produces equipment for the plant and some prototypes. It includes conventional drilling, milling, and boring machines, and some numerically controlled machines. Each machine requires 1 operator.

Units 1, 2, and 3 are engaged in mass-production, whereas the tool-making shop (Unit 4) is small-batch production.

The term "automated unit" refers to the characteristics of machine processing (transfer lines). This does not mean that all workplaces are automated: in these units, we find automated and manual workplaces. We considered the following to be automated workplaces: operators/set-up men working on the transfer machines, having both monitoring and minor maintenance tasks.

In contrast, workplaces with repetitive jobs such as simply loading or unloading the machine were considered as manual. The distinction is also reflected in the training requirements.

In both automated and non-automated production units, maintenance men belong to an independent service department, not included in the study. So we included only those maintenance men with mixed functions: operators/maintenance men who belong to the unit.

In all four departments people work in two shifts: a morning shift (6 a.m. - 2 p.m.) and an afternoon shift (2 p.m. - 10 p.m.).

2.3 The men

All workers interviewed were male and the majority are French citizens from the Monbeliard vicinity. The few foreign workers have also lived there for a long time. People in A units are slightly older than the people in NA units. Length of service in the plant is generally high with about two-thirds of the workers having been with the plant more than five years. 80 %of all the workers, equally distributed in both sectors, have completed primary school. More than half of those in nonautomated units have no technical education and nearly one-third of those in the automated units. In the automated units one-third of the workers, and in the nonautomated units half of them, reported that they received training for their present job. For the majority the duration was less than six months.

3. RESULTS OF THE STUDY

3.1. Introduction

In this part we will report on the results of the study concerning working conditions, work content and environmental circumstances, referring to main parts of the research instruments.

First, we must point out the fact that the interviewed workers did not themselves compare automated and nonautomated jobs, or workplaces, but only evaluated their own workplace and job, regardless of its level of automation. The answers of workers in A units were then compared with those from NA units.

If we consider statistical results obtained by contingency tables comparing automated and nonautomated units, we observe that there are few significant differences with reference to either workplace observations or worker attitudes. "Automation", measured in terms of the technological character of the unit, has little effect.

The research design was later refined with the introduction of work organisation as an intervening variable. In our sample, the influence of work organisation is reflected in differences between workplaces. In comparing A and NA workplaces instead of A and NA units, we found a greater number of significant results.

If we consider this differentiation, we find that the definition of workplaces is not only determined by the technological organisation of machinery, but also by the choice in task division. Moreover, these choices generally lead to a structuring of workplaces according to a hierarchical structure of skills. We will see that this skill structure has many implications for working conditions and work content and in influencing workers' attitudes.

Another characteristic feature of our study which fits in with our previous results on the same topics shows a discrepancy between the opinions and feelings relative to job content and working conditions, and attitudes concerning occupational and social job environment in general. The first aspects are strongly related to the specific features of the job while the second is generally due to ideological and cultural influences. The topics and ideas influencing the working class, as seen by our sample, appear especially through the claims and suggestions developed by trade unions.

3.2. Attitudes towards working conditions

The physical working conditions are generally perceived as rather bad, especially by workers of the A units. At the same time, the importance attached to working conditions is very strong.

In the three production units, the physical working conditions do not vary greatly. In the tool-making unit, on the other hand, conditions are better, the shop is clean, correctly lit, and, in some cases, there is air conditioning. Concerning physical strain, we observed differences between A and NA units arising largely from the number of manual transfers in NA units. There is a discrepancy between these results and the overall attitude related to automation. Fifty four percent of A unit people feel that automation decreases physical strain vs. 20 % feeling that it increases; yet physical working conditions are perceived as worse by people in A units than by those in NA. The less favourable conditions are related to work organisation and the continued high proportion of loading actions, material handling, and manipulation of heavy pieces.

The risk of illness and injuries is not perceived as very great. Nevertheless, the wish to avoid them is very strong.

Workers at only a few workplaces considered time pressure to be excessive. This may be due to the fact that the main types of tasks observed were either routine tasks or long machine sequences without human intervention. Furthermore, in the tool-making unit the precision and quality of work demanded is more important than haste and emergency procedures.

Less than one-fifth of the workers consider their work load as very high. About half of the workers considered that it is fairly high, while the remaining are satisfied. In contrast, 60 % of the workers wish their work load not to be too high. But, generally, this point does not seem to be one of great importance.

The two-shift system is generally considered acceptable by the workers. On the other hand, it may be pointed out that there is a strong sentiment among workers in favour of shorter total hours[3].

The pay structure is primarily related to the skill structure according to conventional scales used for wage calculation and is represented in Figure 1.

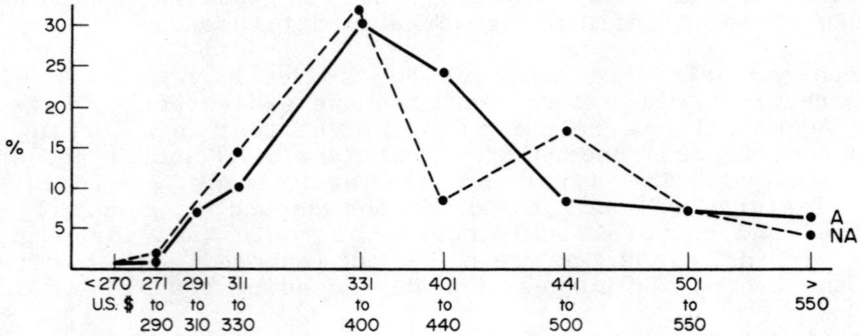

Figure 1. Pay structure

We can consider that the bi-modal character of the NA-curve is related to the differences between workers in unit 2 (fabrication unit, unskilled or semi-skilled workers) and those of unit 4 (skilled workers). The end of the curve corresponds to the wages of foremen and supervisors.

If we take into account the workers' attitudes towards salary, we find that about 10 % rate their pay as being good. About 60 % are somewhat satisfied with their pay, and 30 % consider it insufficient. For about 95 % of the respondents having good pay is considered to be very important.

In spite of the fact that there is a very low rate (about 3 %) of previous unemployment in all units, which is related to the regional characteristics of the plant, job security, is perceived as a very important aspect. At Peugeot, the workers generally consider that they enjoy rather high job security, only 10 or 15 % rate it as very weak. This is in correspondence with all the statements, including those made by unions. Also, considerable labour turnover has been observed in the firm, but for a long time there have been no collective lay-offs. However, there is a risk hanging heavily over foreign workers. It may happen that their employment would not be continued.

Finally, the fear of losing one's job because of automation is more manifested among workers in NA units (37 % vs 17 %). Only one-third of the workers think that a suitable new job would be provided for them. In connection with automation, workers tend to believe that job security will decrease.

There is no personal contact required with co-workers for about two-thirds of the jobs studied. Contact with supervisors is not required for about half of the jobs. So it seems that the general situation is mainly characterised by isolated jobs. However, communication is more difficult in the A units, essentially because of the size of the units, the scattering of the workplaces and the noise. In spite of this, the group relations do not present any particular problem. Observation of workplaces indicates a significant difference between A and NA units related to contacts permitted by the work environment. The noise seems to be more disturbing in production units. In A units communication of machines also occurs by signals.

The reactions to the question concerning relations with supervisors were generally favourable and are summarised in Table 5.

Table 5. Workers' evaluation of immediate supervisor

To what extent does your immediate supervisor:	A			NA		
	Low	Middle	High	Low	Middle	High
give you sufficient explanation?	6.8	37.3	54.2	14.9	19.2	59.5
leave you alone?	5.1	20.4	74.5	10.7	12.8	74.5
listen to you?	0	15.1	84.9	6.4	19.2	72.4
have sufficient skills?	3.4	20.4	76.2	8.5	21.1	70.4

As the table shows, appreciation of the supervisor in most cases is good. In spite of physical difficulties, the social relations between colleagues do not present any particular problem. Contacts between people are numerous and easy.

3.3. Work content

At first, it should be pointed out that automation in the automotive industry does not involve spectacular technological changes. Transfer lines do not imply integrated processes with sophisticated settings. Moreover, transfer machines appeared in the early 50's and today do not look very modern.

On the other hand, there is a change in the number of jobs and their skill requirements with the installation of transfer lines. Except for a small number of set-up men, the remaining workers are unskilled operators, taking care of those tasks which for various reasons were not integrated in the transfer devices. For example, in unit 3 (A), one transfer machine corresponds to 20 machine workplaces and is operated by one worker. In the same unit, six machine-tools require five operators and in addition ten manual workplaces are held by ten operators.

3.3.1. Skill structure and technical requirements

We observed the main differences in division of tasks between the three production units and the tool-making unit. In the former, where there is mass-production, the workers are principally unskilled or semi-skilled operators. The tool-making unit, with small batch production, consists of skilled workers.

Table 6. Distribution of workers in the units according to skills

	Production Units			Tool-making Unit
	1 (A)	2 (NA)	3 (A)	4 (NA)
Semi-skilled workers	17	25	26	
Skilled workers first level (lower)	2	2	0	4
High-skilled workers and others (including foremen and supervisors)	4	6	4	17
Total	23	33	30	21

Table 7. Level of technical education

	Units	
	(%)	(%)
No technical education	28	13
C.A.P. (technical diploma lower level)	58	40
B.E.I. (technical diploma middle level)	11	47

The skill structure is related to job required knowledge and training. The differences are not pronounced between A and NA units, but rather between production units and tool-making units[4]. However, workers in NA units more often think their job requires special training and knowledge. Also wishes for a job requiring higher qualifications are more strongly expressed by the workers in NA units. This is consistent with the higher number of workers in NA units who currently participate in some form of study (9 % vs. 2.4 % in the A units).

At automated workplaces, training is generally acquired on the job. The training time is rather short, frequently less than one month, or in some cases three months to one year, depending on the job. In the nonautomated production unit where the work is manual, it is possible to be trained in a few days. By contrast, as Table 6 shows, the skilled worker in the toolmaking unit requires more technical education.

These differences in job requirements are a function of differences in the work process. This is especially clear concerning surveillance of work process, lifting of heavy pieces, and such aspects of material processing as fastening of the piece, starting of operating process, release and removal of the piece, and control by machine or man. In operating transfer machines, we also find differences in setting up of the machine, attention to detail, positioning of the piece, selection of tool, integration of process and use of special devices such as keyboard panels, optical and acoustical signals, and information listing.

3.3.2. Workers' attitudes toward physical and mental stress

The differences between A and NA workers' perception of physical strain are big: in A units 34 % rate it high against only 19 % in the NA units. On the other hand, low ratings are given by 29 % in A and 43 % in NA units. Reactions to mental stress vary considerably. The workers in A units are more inclined to evaluate mental stress as low (42 %in A, 34 % in NA). To about half of them it is important that a job does not contain mental stress. Actually, observations by experts indicate that mental requirements are not very high and are mainly concentrated in the tool-making unit where it is necessary to classify information and to interpret work drawings. They appear also in the form of visual stress.

If we compare the workers' opinions on stress with the consequences they attribute to automation, we again observe a discrepancy. Automation is seen as increasing mental and nervous strain. 49 % in A units and 40 % in NA say that automation makes mental and nervous strain increase, while 25 % in both cases say that it makes it decrease. This demonstrates a contradiction between direct experience on the job and sterotypes of automated jobs prevalent in the culture and expressed in mass media.

3.3.3. Attitudes toward other characteristics of the job

The most important aspects are variety, repetitiveness, independence, responsibility, and routine tasks.

Significant differences exist between A and NA units concerning repetitiveness of work. It exists in both kinds of units, but is dominant in 17 manual workplaces in A units, i.e. 46 % of the workplaces.

Routine procedures which are more pronounced in the automated units, seem to be linked with the lower level of skill in these units.

The total sample, in general, perceives the job as rather varied. Only one-fourth of the workers in A units, and one-eight in NA units, consider their job has little variety. One-third in both groups rate it as medium, the remainder consider their jobs to have great variety.

Control over workprocess is mainly expressed as "independence". Workers in the A units appear not so tied to a precise place as those in the NA; but, in general, the NA workers feel more independent than those of the A units. This is probably due to the fact that A workplaces are technically rather integrated, in spite of the possibility to move within the unit. On the contrary, in NA workplaces, each operator/machine couple is independent, and the man has the feeling that he is master of his machine.

People in all units consider that their work includes a rather high degree of responsibility (middle: 25 to 30 %; high: 59 to 62 %).

3.3.4. Attitudes related to job abilities and upgrading possibilities

The wish to have an interesting job is general, about 90 % in both units. About half of the workers in both units regard their present job as interesting. The remaining 50 % are equally distributed in two categories: not interesting and rather interesting. The same tendency is reflected in the high importance attached by workers to the possibility of using their knowledge, learning new things, and developing abilities; also to the opportunity for upgrading.

Desire for a job which requires the acquisition of new knowledge is more often expressed by workers in the NA units (A: 65 %; NA 82 %). This may be explained by a higher degree of qualification in the NA unit (tool-making sector) and also by the lesser age of the workers. The least qualified workers in A units seldom aspire to improve their qualifications. If we consider how the present job is estimated to fulfill these wishes, we find that: about half of the workers in A units, and 40 % in NA units, consider that their job does not permit them to develop their abilities. Only about 27 % in A units, and 34 % in NA units, indicate that their jobs offer them any chance for personal development.

As to career opportunities, workers do not feel that chances for promotion are good, that they have much chance to develop their abilities, or that they have an opportunity for more education. Yet all three of these possibilities are considered important and desirable in any job. We found significant differences in possibilities for upgrading, etc. as between skilled and unskilled people, and partly between people working in automated workplaces (semi - skilled) and nonautomated ones (highly skilled).

When considering factors that give a person the best chance to improve his position, the workers ranked them as follows:

(1) Skill
(2) Being a favourite of the supervisor
(3) Working hard
(4) Seniority.

Considering the problem of general satisfaction with work, one has to take into account the content of the work. Only a few people find their job is very satisfying: 12 % in A units, and 9 % in NA units. About half of the workers said they are somewhat satisfied, one - fifth indicated they are partly satisfied, partly not. The distribution of responses shows that general job satisfaction is the same in A and NA units. In both groups, however, more than half said they would like to have a different job. Among these, about 35 % expect they would have a possibility of changing their job, but only 13 % actually intend to change. Correlational analysis indicates that job satisfaction is not directly related to work content and working conditions, but rather to aspirations and values of the workers, as well as other individual characteristics. This shows that general job satisfaction integrates several aspects of the individual's life at work and out of work.

3.4. Attitudes related to automation

Here we review the questions which were put to the workers explicitly dealing with the problem of automation.

In general, the workers seem to feel that management provides insufficient information about the introduction of new equipment. Furthermore, those workers who had been informed said that the information was given less than six months prior to the installation. This may indicate important problems of communication. About half of the workers consider the introduction of new machinery, in general, improves the working conditions in the plant. Only 13 % indicated a negative effect. Almost all the workers were ready to approve and support the introduction of work-saving machinery, while only 10 % would disapprove or resist it. The most important reason for approval by about two-thirds of the workers, is increased efficiency. However, the risk of unemployment worries about 15 % in the NA units.

Table 8. Reasons for approval/disapproval of introduction of new machinery

	A (%)	NA (%)
It saves work	66.1	70.2
It causes unemployment	15.3	8.5
It saves work, but there are risks of unemployment	8.5	4.3

These attitudes, as reflected in Table 9, should be seen against the background of the popular image of automation as a phenomenon freeing man from physical labour and increasing the complexity of the production process.

These results should be compared with the opinions of trade unions concerning the importance and influence of automation. This procedure gives an indication of the level of agreement between the opinions of the union officials and the common workers. First, we will look at the dissimilarity of opinion among the different unions.

Table 9. General consequences of automation*

		Increases	No change	Decreases	Don't know
Job security	A	20	32	37	10
	NA	15	36	45	4
Participation of workers in decision-making	A	19	46	15	20
	NA	0	55	25	19
Prospects for young people	A	20	36	27	17
	NA	17	36	30	17
Physical strain	A	20	24	54	2
	NA	17	28	51	4
Mental and nervous strain	A	49	24	25	2
	NA	40	30	25	4
Job satisfaction	A	25	37	25	12
	NA	28	40	25	6
Importance of unions for workers	A	5	49	10	36
	NA	11	47	0	43

* percentages

Table 10. Opinions of local union leaders on importance of problems of automation

	C.G.T.	C.F.D.T.	C.G.T.-F.O.	C.G.C.
Not important			x	
Somewhat important		x		
Very important	x			x

In general, the C.G.T. declared that the effect of automation on job security depends on whether the savings from automation are used for profit, or improvement of working conditions. For the C.G.T. at Peugeot, automation means a decrease in the number of jobs. For example, the press line was entirely automated and there are now only set-up and maintenance men, and workers unloading the machines.

Table 11. Opinions of local union leaders on specific consequences of automation

Issue	C.G.T.	C.F.D.T.	C.G.T.-F.O.	C.G.C.
Job security	(*)	(*)	Increases	No change
Participation of workers in decision-making	No change	No change	Decreases	Increases or might increase
Prospects for young workers	No change	No response	No change	Might increase
Physical strain	Increases	Decreases	Decreases	Depends on the typpe of machine
Mental strain	Increases	Increases	Decreases	Depends on the type of machine
Job satisfaction	No change	Work is more interesting	Decreases	Increases
Importance of unions for the workers	No change	Don't know	No change	Might increase

(*) for opinions see text

For the C.F.D.T., the number of workers is said to decrease and the jobs require more skill. Therefore, they feel that the problem of unemployment might appear.

The opinion of the C.G.T. leaders towards physical strain - that it increases in A units - is nearer to the workers' perceptions of their work situation. The opinion of the C.F.D.T. is nearer to the workers' general attitudes towards automation which corresponds to the cultural sterotypes of automation.

According to the C.G.T. at Peugeot transfer machines did not introduce big changes in physical strain. Also there may be problems of rhythm. For C.F.D.T. officials, automation introduces less strenuous and more interesting work, requiring higher qualifications.

The C.G.T. thinks that the workers in automated sectors have problems which increase need for assistance by unions, but contacts are often difficult because of the low rate of union membership.

If we compare Tables 9 and 11, we see that the opinions of workers concerning job security are generally in line with those of C.G.T. and C.F.D.T. But an important number also think there is no change, which corresponds with the view of C.G.C.

As to the problem of participation of workers in decisionmaking, the trends are the same. The C.F.D.T. estimates that participation of the workers in decisionmaking depends on top management, and that automation has no influence on this. Considering prospects for young people, the representatives of trade unions appear more optimistic than the interviewed workers.

According to job satisfaction, a greater number of workers express tendencies closer to attitudes of the C.G.T. and C.G.C. The same applies for importance of unions for the workers.

All in all, opinions about consequences of automation are relatively independent of portance of unions for the workers.

All in all, opinions about consequences of automation are relatively independent of the belongingness to A nor NA units but they go often in the way of the attitudes of the most representative trade unions.

3.5. Trade union activities and workers' participation

In general, the orientation of the confederations, C.G.T.-F.O., C.F.T.C., and C.G.C., is to be in favour of participation. The C.G.T. and C.F.D.T., even if they participate in boards and councils, insist on their contesting role and withdraw sometimes from such organisations as a consequence of some problems or conflicts. This is why we must consider the problem of participation in decisionmaking in relation with the problem of trade unions' influence and participation to economic and social life.

Before reporting on this matter we shall insert here figures of the representativity of trade unions at Peugeot plant.

Union membership in the plant amounts to about 15 % of all respondents which globally corresponds to the plant statistics provided by the union offices. This rate is low in comparison to other plants. The rate of membership in NA units is twice that of A units.

Table 12. Representation of trade unions in the sample

	C.G.T.	C.F.D.T.	C.G.T.-F.O.	Affiliation	Total
A	3.4	3.4	1.7	1.7	10.2
NA	8.5	4.3	0	8.5	21.3
TOTAL	5.7	3.8	0.9	4.7	15.1

Table 13. Percentage of votes in the elections

	C.G.T. (%)	C.F.D.T. (%)	C.G.T.-F.O. (%)	C.G.C. (%)
1st college (workers)	62.0	22.43	5.31	-
2nd college ("white collar" and management staff)	19.74	21.32	12.40	30.75

3.5.1. Participation in plant decisionmaking

According to the C.G.T., it is necessary to distinguish clearly between: the participation of workers in decisionmaking (which does not officially exist in France); and the rights of workers, their elected representatives, and the trade union concerning collective bargaining, information and supervision[5].

While there are laws in France regulating occupational life, the C.G.T. does not consider that these laws give the workers the role of participants in decisionmaking. Neither do the employers consider that workers have a right to participate in decisionmaking. Furthermore, people in general perceive an increasing gap between the legal provisions and the way in which the employers apply them. This gap is a source of numerous conflicts which result in the recognition of the importance of social pressure. This is in line with the fact that 15 to 20 % of the workers interviewed believed in strikes and demonstrations as forms of participation in decisionmaking.

According to the opinions of those interviewed at Peugeot, the possibilities for participation in decisionmaking in the plant seem to be very restricted, especially in regard to development of production plan, personnel planning, and disciplinary measures. More opportunities were perceived to exist with respect to introduction of new machinery, determination of incentives and of wage payment, promotion, and selection of workers for training courses. The biggest possibilities were seen in such matters as working conditions and transfer to other workplaces[6].

All the workers declared they were interested in participating in decisionmaking. But their opinions varied with respect to the matters in which they would like to participate. About 22 % in the A units, and 50 % in NA units, consider that workers should have the opportunity to participate in decisionmaking on all matters affecting their plant. Approximately 15 % in A units, and 34 % in NA units, stated that they were personally willing to participate in such decisions.

The largest part of the remaining respondents favour participation only on problems related to their own work (58 % A, and 40 % NA). About 55 % (A), and 57 % (NA) expressed their willingness to participate personally in these matters.

The specific ways by which the workers have tried to influence management are: expressing ideas to the supervisor or management (48 % A and 57 % NA) which is related to the system of suggestion in the plant: expressing ideas to workers' representatives in the plant (in the NA unit 40 %; only 22 % in the A unit). Methods of collective pressure are used by 19 % (NA) and 13 % (A), while for participation in meetings and committees the percentages were 17 % in NA units and 12 % in A units.

3.5.2. Attitudes towards trade unions

In evaluating the data on attitudes concerning the role of trade unions, we need to keep in mind the small number of workers who actually belong to the unions at Peugeot and the fact that even these are divided among a large number of unions. This may help account for the paradox of widespread dissatisfaction with the trade unions, despite convergence between the aspirations of workers and demands pressed by the unions.

Table 14. General satisfaction with the trade unions

All in all, how satisfied are you with the union(s)?	Percentage of workers
Don't know	19
Fully satisfied	5
Somewhat satisfied	31
Partly satisfied/ partly dissatisfied	22
Somewhat dissatisfied	15
Fully dissatisfied	9

Only 36 % of the workers indicated that they were satisfied with the trade unions. The amount of dissatisfaction was the same in A and NA units. In trying to explain these attitudes, we don't know if the dissatisfaction was directed against all the unions or only some particular one. Also, while respondents indicated they were generally interested in the goals the unions pursued (as shown in Table 15) they may have differed on the relative importance of these goals, or the means of achieving them. This part fits in with the lower rate of union membership.

All in all, the workers considered the following union activities as important:

- ensuring job security
- improving wages
- improving working conditions and work content
- developing cultural and leisure activities
- handling workers' grievances and disputes.

National social and economic policy and the workers' political attitudes were seen as less important subjects for union activity. Among all the issues, the four major unions tended to agree that questions of salaries and working conditions were the most important.

Table 15. Issues of highest importance to the union

C.G.T.	C.F.D.T.	C.G.T.-F.O.	C.G.C.
Salaries	Salaries	Employment	Collective provisions
Working conditions (technical and social)	Working conditions	Salaries	Salaries
	Employment	Working conditions	Working conditions
			Specific problems for management personnel

3.6. Commitment to social goals

In connection with broad social issues, the workers were most concerned about:

- war and peace
- health care
- pollution
- youth problem
- housing problems

We can say that there is no problem that does not interest the majority of the workers; although the commitment of the workers in the NA unit seems to be less strong. If we consider the extent of actual participation in solving these problems, we find that it is not big. About one-third of the workers reported that they had been active in the solution of at least one of these questions.

The means most used by workers to exert influence are talking with union representatives and with party officials, and participating in strikes and demonstrations. So, we see the importance that workers attribute to means proposed by the unions and to the struggle with management through trade union activities. On the other hand, "to work actively in committees" is seen as the most official means of participation and hence discounted.

Table 16. Participation in activities of organisation (ranked)

Automated units	Percentage of yes responses	Nonautomated units	
Sport organisations	40.7	Sport organisations	53.2
Municipal organisations) Educational organisations)	25.4	Youth organisations	27.7
Trade unions) Youth organisations) Cultural organisations)	18.6	Educational organisations	23.0
		Municipal organisations	21.3
Social organisations	15.3	Trade unions) Cultural organisations) Social organisations)	14.9
Political parties	10.2	Religious organisations	12.8
Religious organisations	6.8	Policitical parties	10.6
Organisations maintaining public order	0	Organisations maintaining public order	0

Turning to organisational activity, it seems that interviewed workers mean that they support the organisations named, rather than really taking part in their activities. For example, the answers concerning the trade unions indicate that the number of people who declare that they participate in the activities of the trade unions is higher than the numer of people belonging to the unions. Moreover, the proportion of workers who affirm that they participate in trade union activities, respectively 18.6 % in A and 14.9 % in NA sectors, is reversed compared with the number of people belonging to the unions.

Obviously these responses concerning commitment to social goals and participation in activities of organisations cannot be considered coming as they do from a single conclusive plant. But we consider it important to mention them for comparison with other countries.

4. CONCLUSIONS

It is clear that the rise of automation leads French workers, in general, to express fears about unemployment and changes in work content. People demand jobs of a higher skill level and relevant training. In connection with this specific plant, though, the workers are not too preoccupied with the danger of unemployment because this problem has been solved here. This results from collective pressures and actions by the trade unions.

In general, the comparison of automated with non-automated units in this specific plant does not show significant differences in workers' attitudes. This may be due to the fact that the introduction of automation in the French automotove industry is not recent. Therefore, many of the interviewed workers in the automated units have never experienced other ways of doing their job. Also, the equipment represents a rather low level of automation. The investigated technology can be characterised more as "automatic" than an example of line automation. There are, for example, no integrated computers nor a remote control of the process. Other problems like internal transfers are probably seen as more important by the workers than technological changes. As Pierre Naville said in "Vers l'automatisme social?", alterations always take place in industry. Such minor changes often have more direct impact on the level of units and workplaces.

Similarities both between units and workplaces result from the specific role of work organisation which reproduces the traditional division of labour and hierarchical skill structure in the automated units. Taking into account these characteristics we can still notice some differences between automated and nonautomated workplaces. In automated units we find people working at new workplaces or at traditional and manual workplaces with simple machines and low skill demands. In nonautomated department there are two subgroups: low and high skilled workers. This difference in skill level between automated and nonautomated units does cause differences in workers' attitudes.

However, it is difficult to isolate the impact of skill structure from the effect of better conditions enjoyed by more skilled workers. At a macro-sociological level both groups belong to the working class. But in a micro-sociological perspective, we must consider differences between the two groups in terms of training, wages, promotion and upgrading possibilities. The skilled workers are also more often members of the trade union which implies more activistic behaviour, but also that they are better informed.

All in all, we cannot conclude that there is a strong correlation between technological change and attitudes of workers. Nor does the study invalidate the general consensus of the French working class struggling against the employers, in spite of the plurality of trade unions and the specific climate of each plant.

Also the trade union policy has a homogenising effect on the workers attitudes which diminishes the differentiating impact of differences in technology.

NOTES

1 We may refer to the role of the "Automation Committee" dealing with relations between research and industry in the "General Delegation for the Scientific and Technical Research".

2 Le Monde. June 10, 1976, p. 19 "Diffusion et creativite a Montbeliard", by Marie-Francoise Levy.

3 For many years trade unions in France have demanded a standard 40 hour week, instead of 43-46 actually prevailing.

4 This difference comes also from the fact that the maintenance men (more skilled than the operators working on the transfer lines) do not belong to the A units but are grouped in an isolated maintenance unit which we were not able to study.

5 See L. Chavrot, Communication to the colloquium about participation of workers in decisionmaking, ILO, Oslo, August 20-30, 1974.

6 Legally, the Enterprise Committee (including the top manager of the plant and the workers' elected representatives) have to be informed on matters of collective dismissals and collective transfer of workers. But, very often a dismissal is avoided more efficiently by strikes or collective pressure than by the opposition of the elected people in this committee.

4. Polish Automobile Workers and Automation*

ADAM SARAPATA

INTRODUCTORY REMARKS

One of the leading ideas of the international comparative study entitled Automation and Industrial Workers has been to get answers to two intriguing questions:

(1) Does automation exert an influence on the industrial worker's situation in his work and on some selected aspects of his life and if so, to what degree?

(2) Does the social, economic, and political system affect the process and outcome of automation in the sphere of industrial work, and how?

The controversial hypotheses, views, and results on the outcome of technological development in the field of workers attitudes toward work and life can be summarised in the following way:

* The research in Poland and the preparation of this paper could be done only due to the friendly assistance of the Ministry of Machine Industry, the Central Council of Trade Unions, and the Management of the Institute for Organisation of Machine Industry.

I would like to express my great thanks to these institutions, to the managers of the investigated enterprises, to the anonymous workers who truly responded to the questions of a rather long questionnaire, and, naturally, to all friends and colleagues - the participants of this international study - for their assistance and co-operation in this exceptional venture - the project on Automation and Industrial Workers.

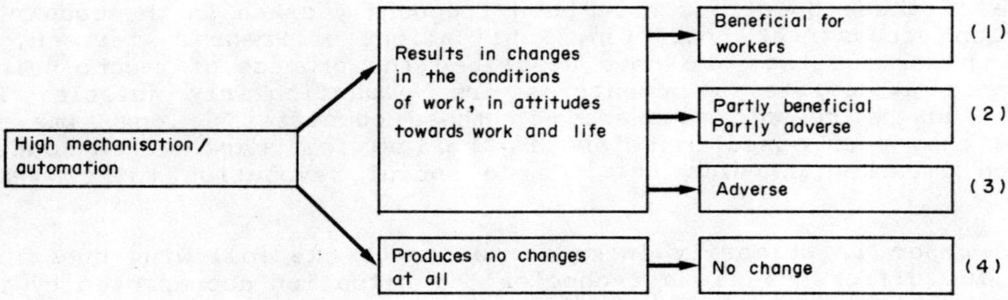

Some research workers and authors claim that technological advance and automation work in a universal way to produce results of a similar kind, regardless of the socio-economic system and other specific features of the country. Others maintain that systemic features modify the process of technological development, the situation of workers, and their attitudes towards work: for instance, within one political system technological advance may be beneficial and a real blessing while in another its effects are adverse and it becomes a destructive force that threatens the whole society. According to some other claims technological advance does not substantially affect situations, attitudes, and the behaviour of workers as well as the given sphere of their lives.

The material collected in Poland, obviously, does not alone provide grounds to answer the questions that can be cleared only by comparative analysis of the data coming from countries of various social and political systems. Nevertheless, the Polish material, as well as the material collected in other countries, contributes to the answers to such essential questions as: what kind of psychological and social effects arise within the same global, social, economic and political system, but under different technological conditions? Does automation affect the situation of the industrial worker in his work and, if so, to what degree? Which of the four situations presented in the above diagram will prove most common in the plants of a socialist country? Will it be only one of these situations?

In this report I shall concentrate on the following questions:

(1) What importance do workers employed in nonautomated workplaces (NA workers) and workers employed in automated workplaces (A workers) attach to various aspects of work?
(2) How do NA and A workers evaluate the current state of individual components of their work?
(3) What makes one generally satisfied? What are the components of overall satisfaction? What are the correlates of satisfaction? How satisfied are the workers who man technologically different workplaces? Does handling of the technologically "higher" workplaces bring higher satisfaction?
(4) How do NA and A workers feel about technological change? Are both categories of workers equally willing to support the installment of new equipment and new machines? What do the

respondents expect from automation? What do the attitudes of the respondents toward technological advance depend on?

Some of these questions have been frequently asked in the countries of long industrial tradition. Publications of French, American, or German sociologists who have described the process of technological advance in capitalist countries are particularly notable. The questions which were answered in those countries a long time ago have only been posed in Poland in the last few years - in a country which embarked on industrial and social revolution thirty years ago.

This paper is primarily intended to answer the following question: Are the differences in a technological situation accompanied by the same or different reactions (views, attitudes, behaviour) of the workers within the same system?

It would be a real misunderstanding to draw general conclusions that would apply without any reservations or limitations to the whole country or even to the Polish machine industry on the basis of the findings presented here from the three enterprises selected for investigation.

1. CHARACTERISTICS OF THE SAMPLE

The results of the studies presented in this report are based on responses obtained from 206 workers, namely 109 workers manning 49 NA machines (conventional) and 97 workers manning 27 A machines (transfer lines).[1]

The studies in Poland were carried out in three automotive industry plants in A and NA departments. The studies were conducted by industrial sociologists and psychologists in May and June 1974.

Tables 1 and 2 contain data on the age and monthly wages of the sample. Only operators (men) were selected and interviewed. The average length of employment in the given plant was 12.8 years for workers in NA workplaces and 11.4 years for A ones.

Table 1. The age of the workers (a)

Age	(Up to 20)	21-25	26-30	31-40	41-50	51-60	(over 60)	Total
Workplaces:								
NA	3	22	18	36	17	4	-	100
A	8	14	20	36	18	4	-	100

(a) In percentage of a group.

Table 2. Monthly wages in zlotys of the interviewed workers (a)

Wages	(up to 1500)	1501-2000	2001-2500	2501-3000	3001-3500	3501-4000	4001-4500	4501-5000	(over 5000)
Workplaces:									
NA	-	4	10	30	19	24	6	4	3
A	-	5	15	39	21	12	4	3	1

(a) In percentage of a group.

Figure 1 presents profiles based on the average for both categories of workers. It shows that more significant differences between NA and A workplaces concern six activities. The presented data indicate that the selection of workplaces for the investigation was quite correctly accomplished, and the division into these two categories may be used in further analysis and description. The comparison of the criterion of level of automation adopted here in the analysis with other criteria provided results which proved the correctness of our classification of workplaces and workers. (Correlations being about rho = 0.60.[2])

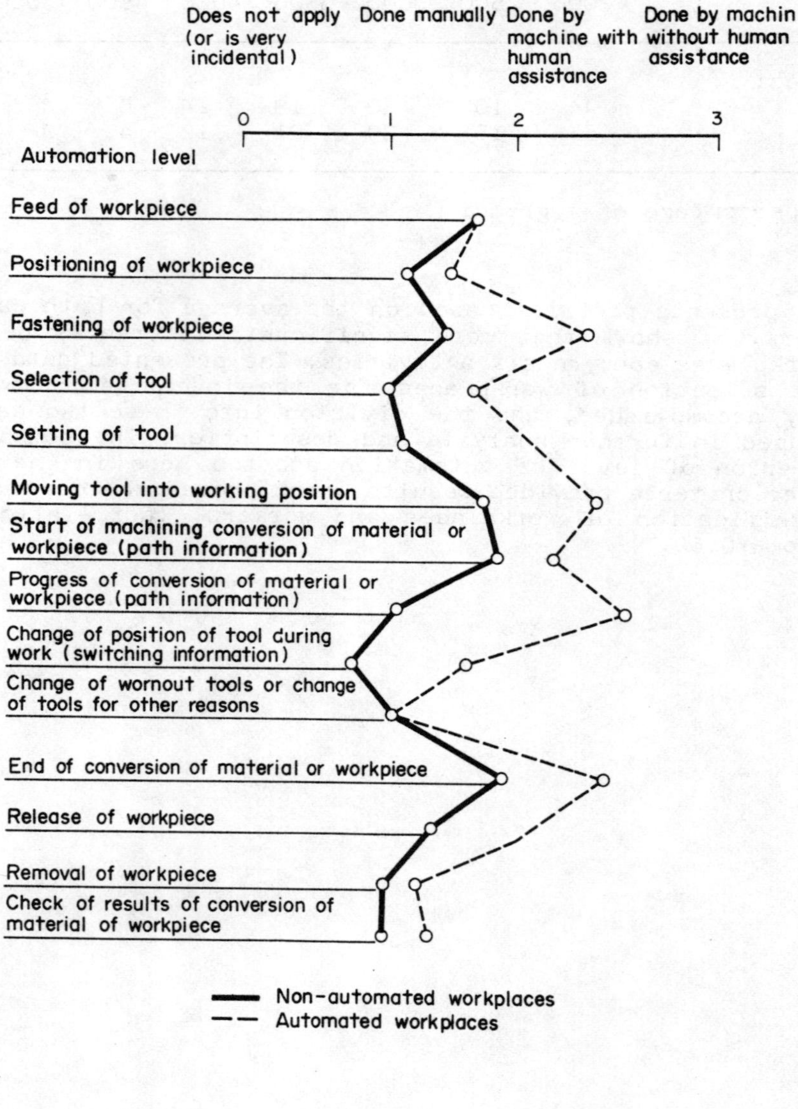

Fig. 1 The level of automation: manual and automated activities at the workplace.

2. ASPIRATIONS AND VALUES OF NA AND A WORKERS

The questionnaire for workers contains thirty one questions which may be considered as questions revealing aspirations and values of the respondents. The questions present a rich set of the workers' needs and aspirations concerning both the realm of their work as well as non-occupational spheres.

It shows that 24 of the 27 elements of the work situation measured with a nine-point scale rating were rated between 7.0 and 8.5. Assuming that ratings 7-9 represent a high degree of importance, a large majority of the evaluated elements of work elicit keen interest among Polish workers.

The results show that Polish workers attach great and very great importance to (the ratings are given in parenthesis):

(a) The content of work:
- it should provide opportunities to improve one's skills (7.9).
- it should be interesting (7.6).
- it should provide opportunities to develop one's abilities (7.4).
- it should be independent (7.4).
- it should provide opportunities to improve one's general education (7.2).
- it should offer opportunities to use one's knowledge and skills (7.2).
- it should be creative (6.8) and diversified (7.0).

(b) The physical conditions of work:
- safe and healthy conditions (8.2).
- good physical conditions of work (7.9).
- within suitable working hours (7.9).

(c) Job security (8.3)

(d) Rewards for work:
- it should be well paid (8.5).

(e) Job prospects:
- it should provide opportunities for promotion (8.2).

(f) Interpersonal relations:
- co-workers' personal interest (8.5).
- helpful co-workers (7.9).

(g) Professional and moral qualifications of the immediate superiors:
- supervisor sees workers are justly rewarded (8.2)
- supervisor gives sufficient instructions (7.8)

- supervisor listens to workers (7.9)
- supervisor is competent (7.9).

Very high ratings on the importance scale of "earnings", "supervisor fairness", and also partly "opportunities for promotion" should be understood in connection with three facts: the material well being of people of the country was ruined by war and Nazi occupation; the immense sacrifices of the nation to build a new system and to industrialise the country; and, finally, the fact that the job is the main source of income, hence the job and earnings (not capitalisation) evoke particularly vivid interest among Polish workers.

A high rating of "job security" on the importance scale is not related to the fear of unemployment. It expresses the need to maintain the work rhythm and to avoid disturbances in production and the desire to be the author of a change of job, profession, or plant. Polish workers are not afraid of unemployment and many sociological findings prove this. Three workers, that is 1.5 %, gave an affirmative answer to the question posed in our studies whether the respondent had been out of work during the last 5 years. In the Polish version of the questionnaire the question: Have you been unemployed during the last 5 years? was translated as Have you been out of work during the last 5 years? The question worded this way in Polish means both a situation of unemployment and, more frequently, a situation of occupational inactivity due, for instance, to change of residence, illness, or work not considered as permanent within the nationalised sector.

Polish workers also attach very great importance to both the conditions of work, such as safe and convenient physical conditions, suitable working hours, and the rewards for work (earnings, promotion opportunities, specifically perceived job security), as well as to the fact that the work should be challenging and provide opportunities to improve one's skills. Unfortunately, the questionnaire did not contain questions that would enable one to grasp various aspects of willingness to "give" on the part of the workers, that is their readiness to contribute and not only to get rewards and make demands.

The findings presented so far cover the sample as a whole. The aspirations of NA workers are similar to those of A workers, and statistically valid differences occurred in 4 cases out of 27:

	NA	A
It is important that the immediate supervisors have sufficient skills	8.2	7.6
It is important that you and other workers from your department help each other at work	8.2	7.6
It is important that you and other workers from your department be friendly with each other	8.6	8.3
It is important that the work provides opportunities to develop your abilities	7.1	7.8

The findings presented in this section indicate that workers employed at technologically different positions have in general "similar" aspirations. I have used quotation marks here because some reservations can be raised as to the same meaning of the same numerical values.

Besides the twenty-seven work components mentioned above, the questionnaire contained several other questions to measure other aspirations and values cherished by the respondents. Two such questions concern the workers' participation in management.

Perhaps it is worthwhile to keep in mind that activists and research workers argue endlessly about what the workers desire: Are they interested in participation in management and, if so, to what degree? Robert Blauner, the author of an American classic on alieniation and freedom, has advanced a theory that workers do not strive for influence in the realm of ownership of means of production hey do not experience alienation because they are not owners of means of production; "ownership powerlessness" is not the source of their alienation. According to Blauner:

> "The average worker does not want the responsibility for such decisions as what, for whom, and how much to produce; how to design the product; what machinery to buy; how to distribute jobs, or how to organise the flow of work . . . as labour reform movements became more sophisticated, they realised that large - scale organisations cannot be governed directly and en masse. The sponsors of direct democracy gave way to the advocates of representative democracy and participation in management. The most important recent examples of this trend are "joint consultation", in England, co-determination in Germany, and the workers' councils in Eastern Europe. Yet the experience of these representative systems suggests that it is only the delegate or the participator, not the average worker, who actually feels he is influencing major decisions."[3].

The above mentioned statements and the general discussion on the participation of workers in management bring up several basic questions: What decisions do the workers want to make? What participation aspirations do workers of capitalist enterprises have and how do workers of socialist enterprises approach this problem? Does automation affect these aspirations?

The cross-national comparison will try to answer these crucial questions in detail. Here only the findings of Polish studies can be presented and some answers can be briefly touched upon.

The question and findings concerning worker participation in making decisions are included in Tables 3 and 4.

Table 3. The workers' opinion about the opportunities they should have in participation in decision making

Question: Now we would like to know your opinion about the participation of workers in decision making at the plant. First, which of the following statements best reflects your opinion about the opportunity workers should have? (Check one.)

Statements (opinions)	(%)
Workers should have the opportunity to participate in making decisions on all matters affecting the plant	48
Workers should have the opportunity to participate in making decisions only on problems affecting their work	45
Workers should have the opportunity to participate in decisions only when management asks them	7
Workers need not have the opportunity to participate in decision making	-

Table 4. The workers' personal willingness to participate in decision making

Question: How would you like personally to participate in decision making in your plant? (a)

Statements (opinions)	Responses (%)
I would like to participate in most decisions affecting the plant	32
I would like to participate in decisions only on questions affecting my work	57
I would like to participate in decisions only when management asks me	9
I don't feel like participating in making decisions	2

(a) The original statement "I am willing" was translated into Polish as "I would like." Thus as phrased in Polish the statement conveyes more attachment, engagement, etc., as a result lowering the percentage of positive responses.

None of the workers is of the opinion that workers need not have the opportunity to participate in decision making.

Ninety-nine % of the workers think that workers should have opportunities to participate in making decisions concerning their work and 85 % express the desire for personal involvement.

44 % of the workers think that workers should have opportunities to participate in making decisions concerning all problems of the plant and 30 % express their personal aspirations in this field.

There is a very high correlation (rho = 0.62) between the views on "workers' right to participate in making decisions" and the willingness to be personally involved in the process. Table 5 shows responses jointly on these two questions.

Table 5. Joint responses on questions concerning the workers' opportunity to participate in decision making and their willingness to participate *

Question: What opportunities should workers have to participate in decision making at the plant?	Question: How would you like to participate in decision making in your plant?		
	Statements		
Statements	Only when management asks me (%)	Only on questions affecting my work (%)	In most decisions affecting the plar (%)
Only when management asks them	64	28	-
Only in decisions affecting their work	4	90	4
In most decisions affecting the plant	4	30	64

* Omitted "Don't know" answers.

The responses to the questions concerning personal involvement and strong connections between the responses to both questions undermine the claim that workers are not interested in matters outside their own work. A conclusion of this kind seems entirely substantiated if one takes into consideration that the kind of work being performed makes it difficult to participate in meetings and discussions during working hours. Many of the respondents live far from their plants, commute to work by plant buses, or are in a hurry to catch other means of mass transportation; thus they have little or no opportunities to meet outside working hours.

Aspirations for participation and the willingness to be personally involved were equally distributed between NA and A workers.

The next question to be presented here concerns worker aspirations as far as the activity of trade unions is concerned. The workers' views on: How important in the activity of trade unions should the following matters be? are presented in Table 6 in descending order according to how important these matters are.

The last question concerning aspirations and cherished values deals with non-occupational civic interests of the workers. For the sake of brevity I shall not present full details here. Let me mention, however, the fact that out of ten evaluated issues the following three were most frequently labeled as "I am very deeply concerned in these issues":

prevention of wars and maintenance of peace	76 %
health protection	75 %
housing problems	69 %

The prevention of wars and the maintenance of peace has been the most essential issue for the Polish nation for a long time. This has been proved by other sociological studies. The interest in health protection and housing problems shows both high aspirations and the existing needs.

3. ASPIRATIONS AND THE EVALUATION OF THE PRESENT WORK SITUATION

While average values attached to twenty-seven individual work components on the importance scale fall within the range of 7.8-8.5, the average values of the present state are scattered all over the scale and fall within the 3.3-7.7 range. Five evaluated components received 7 or higher ratings; the majority took the middle of the scale between 4 and 6.

Aspirations and values of NA and A workers show great similarity with statistically valid differences occurring only in 4 cases out of 27. However, the present state evaluations show more differences. Statistically valid differences concern 13 out of 27 components, that is about 50 %. In 12 cases out of 13 the ratings are more favourable among NA workers.

Table 6. The workers' opinions on the scope of activitiy of trade unions

	Very important (%)	Important (%)	Unimportant (%)
1. Improving wages, pensions, etc., of the workers	89	10	1
2. Developing cultural and educational activities in the plant	88	12	-
3. Providing recreational facilities (including vacations)	82	17	1
4. Improving productivity	78	22	-
5. Awakening a strong political attitude and consciousness among the workers	76	22	2
6. Improving the terms and conditions of work	73	27	-
7. Influencing the general social and economic policy of the country	67	27	6
8. Handling worker grievances (complaints and disputes)	65	33	2
9. Ensuring the job security for workers	51	46	3
10. Changing work organisation to increase job satisfaction (humanisation of work)	49	44	7

The method applied in our study does not guarantee a direct response to the question: To what extent are the respondents satisfied or dissatisfied with individual components of work? We can learn from the questions what importance the respondents attach to specific components and how they see the present state of a given aspect of work. There are no questions, however, concerning satisfaction with these components.

Combination of responses to the questions concerning the importance and the present state may bring one closer to the satisfaction problem but does not solve it.

In Fig. 2 the components of work are placed according to the diminishing difference between the importance and the present state level. The figure contains only twenty components.[4] The figure indicates that:

(1) The extent level of the work components shows much larger variation than the importance level.
(2) The discrepancies between the importance and the extent state are particularly great with regard to earnings, upgrading or promotion chances, physical conditions of work, opportunities to improve one's skills.
(3) Slight differences, not exceeding one degree of the nine-degree scale, concern the relations with co-workers.

The discrepancies between the present state and the importance attached to pay as well as upgrading and promotion chances, which in most cases are conceived as an increase in earnings, are understandable in view of the situation of a country that is striving for a higher standard of living.

The discrepancy concerning the opportunities to improve one's skills results not so much from the lack of such possibilities in the plant but rather from the respondents' high aspirations on the one hand and practical difficulties on the other. An indicator of the possibilities is the observation that more than one-third of the respondents completed vocational training in the plants under investigation and more than 12 % were presently undergoing training.

Such conditions of work as illumination, temperature, cleanliness, and noise level, were given ratings of 4.9 by our respondents. Their importance - 8.2. Experts gave the working conditions more favourable ratings. The reasons for such discrepancies between the evaluation of the workers and that by experts require some explanations. The fact remains, however, that the working conditions should be improved. It is worth mentioning here that the improvement of working conditions has been given top priority in the latest party, trade union, and plant administration programmes.

While studying Fig. 2, it is worthwhile to keep in mind that the joint outcome of importance minus present state occurs in the situation when one of these two components is bigger and not only in those situations when the value of importance predominates. Measurement of this kind seems to evoke no reservations under such circumstances as, for instance, when the respondent rates the present state of "responsibility in his work" as 5, while giving the "importance of responsibility" the rating of 8 or even 3. In the first case one can assume that the respondent wants greater responsibility and hence suffers some kind of unpleasantness (deprivation); in the second case he also experiences unpleasantness because of excessive responsibility, and so deprivation also takes place.

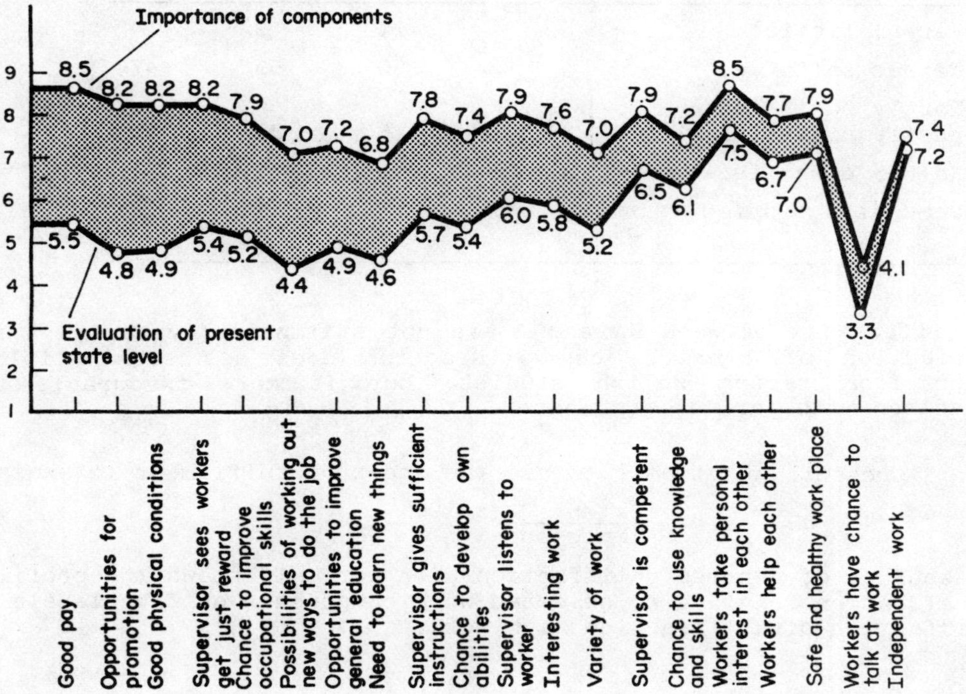

Fig. 2 Distances between importance and evaluation of level of selected work components.

Can each situation, however, be interpreted this way? Let us, for instance, take the respondents who rate their current opportunities to use their skills or promotion chances as 6 and the importance of such possibilities as 4. Are they being deprived?

The above-mentioned situation is real. This, among other things, signals a need to interpret the index of deprivation more cautiously and to look for other ways of presenting the findings in keeping with logic and respondents' intention.

4. GENERAL SATISFACTION WITH WORK AND ITS CORRELATES

The question: How much, generally speaking, are you satisfied with your work? produced the results shown in Table 7.

Table 7. General satisfaction with work

	Total %	NA %	A %
Very satisfied	11	9	14
Rather satisfied	50	53	47
Partly satisfied, partly dissatisfied	28	30	26
Rather dissatisfied	9	7	11
Very dissatisfied	2	1	2

The differences between NA and A are not statistically valid. This distribution of overall job satisfaction does not substantially differ from recent Polish studies, but is more favourable in comparison with earlier findings.

Fig. 3 presents an attempt to use the index of fulfilment to explain

$$\frac{\text{extent}}{\text{importance}} \times 100$$

the sources of general satisfaction with work. It shows the profiles of saturation relating to selected work components on five levels of satisfaction-dissatisfaction with work.

Generally speaking, the higher the level to which needs are met, the greater the fulfilment of expectations, and the higher the level of general satisfaction with work.

In some cases general satisfaction with work increases with the fulfilment of a given component; this holds true with such components as, for example, the immediate superior sees to it that the work done be properly rewarded (Eta = 0.43); proper pay (Eta = 0.36); the work is interesting (Eta = 0.34).

In relation to friendliness among co-workers, fulfilment is stable and at the same time high regardless of the level of general satisfaction (Eta = 0.04).

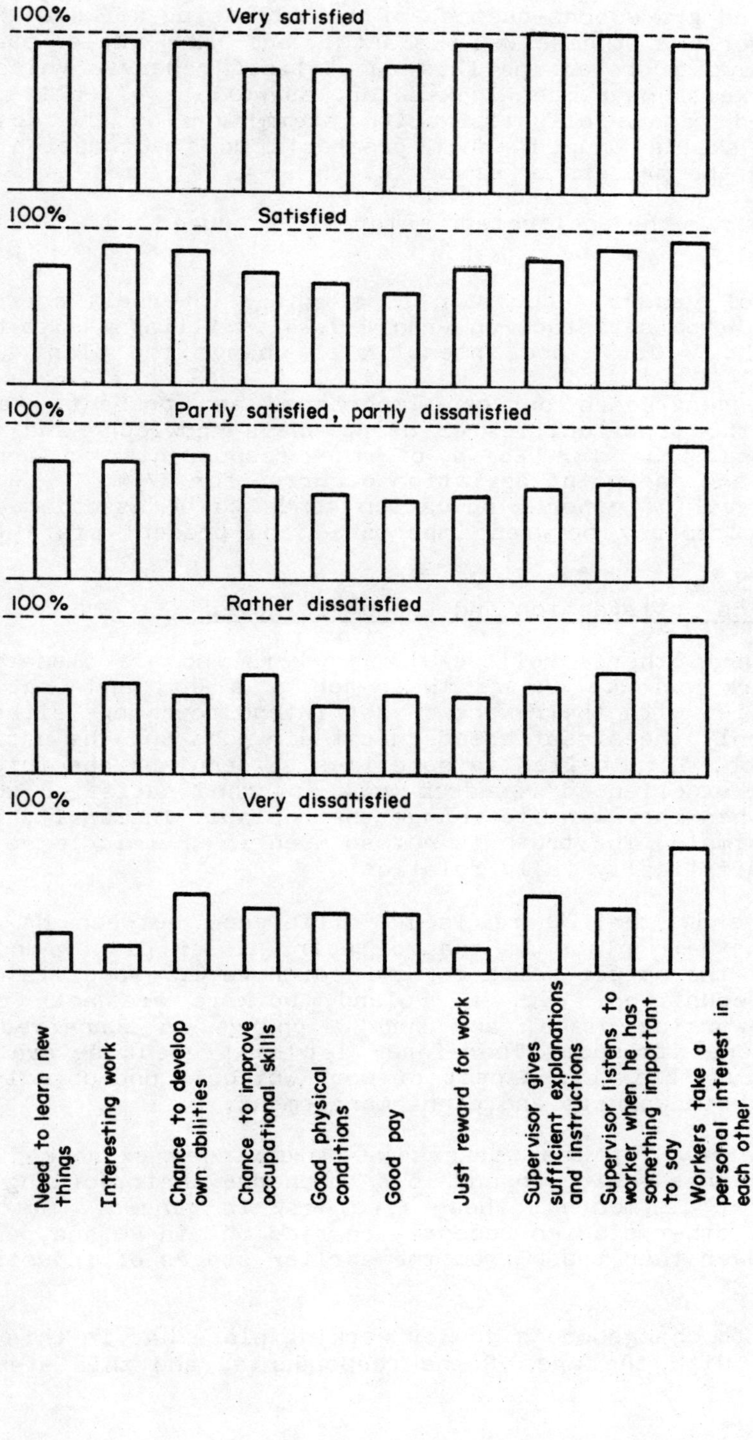

Fig. 3 Level of fulfilment of selected work components (in percentage in five levels of satisfaction with work).

5. GENERAL JOB SATISFACTION AND THE OPPORTUNITIES FOR THE USE OF ONE'S POTENTIAL

More and more emphasis has recently been placed in literature and practice on the extended use of workers' knowledge and skills. The heavy losses and grave consequences of not utilising workers' skills affect the workers themselves, plants, and the whole society. Reasons and ways of releasing these unutilised reserves (hidden in "untapped" workers) have been looked for, as well as effective means to prevent and remove dissatisfaction stemming from the sense of having one's skills unused. What prospects does automation hold? Does it deepen the process or erase it?

The analysis of the collected material produced the following findings:

- the level of general education is slightly inversely correlated with general job satisfaction (rho = -0.14), willingness to change the job (rho = - 0.15) and intention to change the plant (rho = -0.14).
- A complete congruence between importance and present state in regard to the opportunities of using one's knowledge and skills at work was true for 25 % of the respondents; a complete conformity and one-point deviation occurred for 57 %.
- A higher level of general education tends to be associated with greater discrepancy between importance and present state (rho = 0.20).

5.1. General Job Satisfaction and Labour Turnover

Many laymen (and others) believe that workers who are dissatisfied with their work want to change their job or plant, and that those who are satisfied with their work do not intend to change either. In other words, only the dissatisfied quit their jobs and the satisfied stay on the job. This belief is sometimes so deep and absolute that it makes less experienced research workers brand satisfied workers as illogical when they want to change their plant. In reality things are not that simple. The truth is not so much in the middle as above or below a statistically valid relation.

The findings show that there is no difference between NA and A workers as to their plans to change their job or plant, and with this statement the matter could be dropped. However, some claim that in socialist countries, e.g. in Poland, workers are not free to change their working place, and such a change is conceived as a token of personal freedom. Therefore, let me present several data that would illustrate this aspect of work which is undoubtedly very essential for both workers and plant management.

Forty-eight percent of the workers we interviewed expressed their intention to change their job and 15 % to change their working place within the next 12 months. These findings, in general, match the conclusions of other studies recently carried out in Poland, and the figures are lower than those from the earlier stages of industrialisation.

The intention to change one's job or working place has in this study nothing to do with the age of the respondents, and this stands to reason if you take into consideration the fact that one of the criteria for selecting respondents was at least one-year-long employment in the plant; and young workers employed for less than 6 months are most mobile. Beyond that period labour turnover rates show no striking differences between age groups.

Pay shows a certain connection with the intention to change job or working place, but two things merit attention:

(a) There is a stronger connection between the evaluation of pay and the intention to change one's job and working place. The evaluation becomes a filter and that is a decisive factor in making plans to change job or plant. Here are the coefficients:

Actual pay level	wish to change the job	rho = -0.020
	wish to change the plant	rho = -0.221
Evaluation of pay level	wish to change the job	rho = -0.200
	wish to change the plant	rho = -0.281

(b) There was a strong connection between general satisfaction with work and the intention to change job (rho = -0.3) and in particular the intention to change plant (rho = -0.45) (Fig. 4). Among those who want to change their job, 29 % also want to change their working place (Table 8).

Table 8. Correlation between intentions to change job and plant

	Wish to change the plant			
	No %	Don't know %	Yes %	Total %
Want to change the job	55	16	29	100
Do not want to change the job	86	7	7	100

The analysis shows that the change-oriented workers constitute two different categories that differ, among other things, according to pay level.

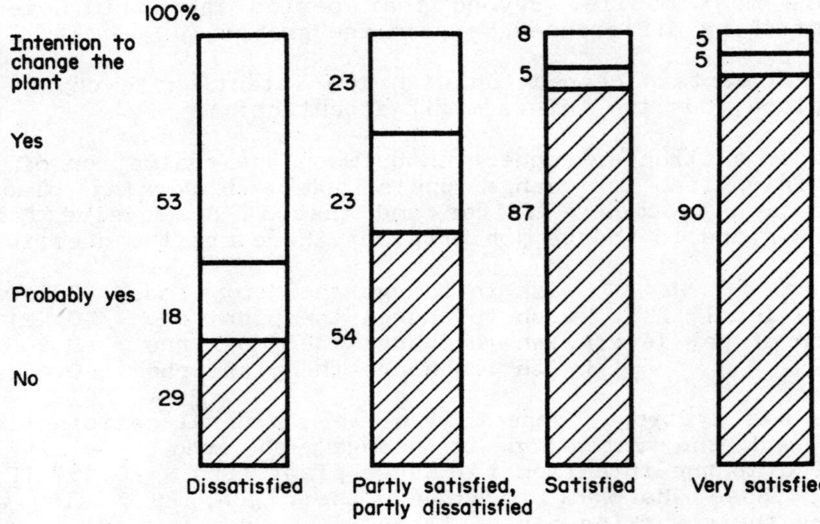

Fig. 4 General satisfaction with work and intention to change the plant within the next 12 months.

5.2. The Problem of Alienation of Young Workers

Many western studies bring up the fact that young workers are in many respects in an inferior situation than their older colleagues, that they are more alienated and hence, among other things, are more willing to change their job or plant frequently. What does the situation look like in Polish workplaces?

Let us begin with the findings concerning the widely publicised sphere of alienation: the use of knowledge and skills at work by young workers. It has been stated in the West that underemployment is particularly common among young workers.

The study conducted among NA and A workers produced the following findings:

The situation that "importance" was rated greater than the "present state" concerning possibilities of using one's knowledge and skills at work, applied to 61 % of workers up to the age of 25, 64 % between the ages of 26 and 30, and 50 % for other age groups.

The course of the curves in Fig. 5 shows the number of workers in various groups with various discrepancey levels between the ratings of "importance" and "present state". These figures and correlation coefficients do not seem to confirm too convincingly the alienation phenomenon among young workers observed in other countries.[5]

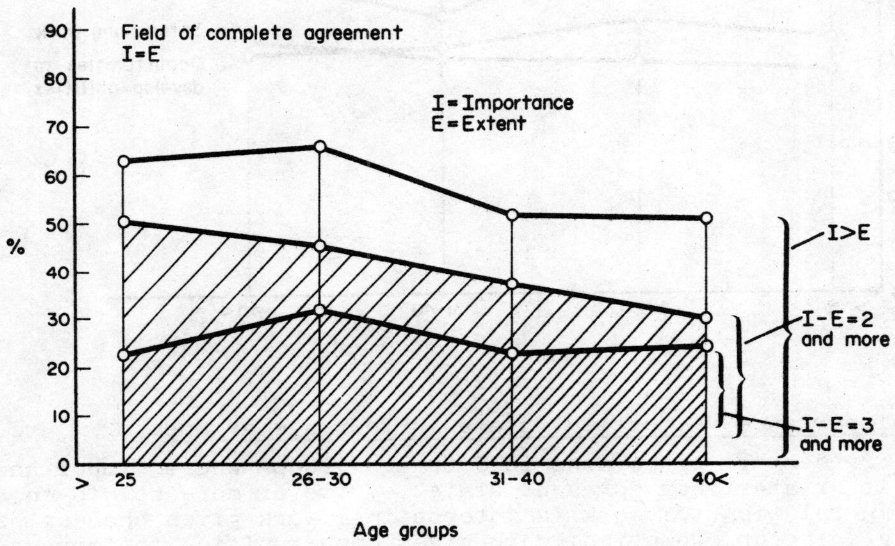

Fig. 5 Opportunities to use knowledge and skills according to age groups.

(Common occurrence of situations with various gaps between importance and present state concerning opportunities to use knowledge and skills at work according to age groups.)

The situation of young Polish workers has not proved worse than that of older workers in many other respects either. The correlation coefficients do not exceed 0.1. Young workers evaluate their work the same way the older ones do. This is true for such work aspects as shown in Fig. 6: the work is interesting, it provides opportunities to develop one's abilities, it offers chances of upgrading or promotion, it provides opportunities to improve one's skills.

Consequently, young workers, despite their aspirations and some impatience, are only slightly less satisfied with their work than older workers. The rho coefficient for age and general satisfaction with work is 0.155.

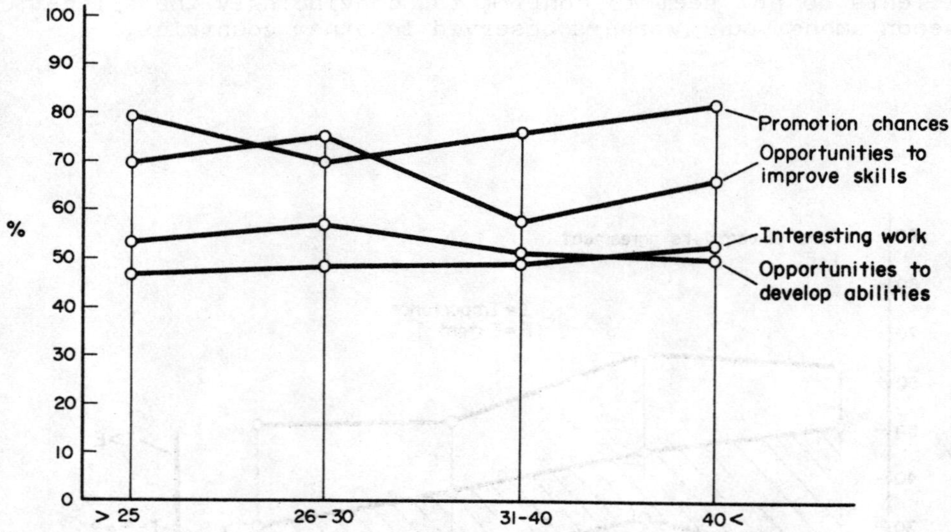

Fig. 6 Importance vs. present state in four age groups.
(Percentage of persons in four age groups where "importance" is greater than "present state" by two or more points on scale in relation to: work is interesting; work gives chances of upgrading or promotion; work gives opportunities to improve skills

Young workers in comparison with older workers are not more eager to change the job or plant:

Age and willingness to change job	rho = -0.02
Age and intention to change plant within the next 12 months	rho = -0.04

6. ATTITUDES TOWARD TECHNOLOGICAL DEVELOPMENT

The epoch-making importance of automation makes the problem of workers' attitudes toward automation theoretically interesting and practically significant. What effects of automation do the workers envisage? Do they see it as a blessing or menace? What does it bring them in their work?

Renowned publications of West European sociologists contain many findings and statements concerning automation to the effect that many workers are afraid of technological advance, new equipment, and automation.[6] American studies offer similar findings.

How do the workers of a socialist country view and evaluate the effects of technological advance, the introduction of new machines and equipment, and automation? Will they react negatively, with reservations, and even with hatred?

The first study on worker attitudes towards technological advance was carried out in Poland in 1959, that is almost 20 years ago.[7] The study covered a representative sample of Polish workers (769 persons). To the question If it were possible to use new machines in a plant which would increase the production and facilitate the work, but which would lay off one-fourth of the plant's crew - what do you think has to be done? the majority of interviewed workers were in favour of installing new machines with the one provision that those laid off should get new jobs.

On the present project the questionnaire contained three questions dealing with worker attitudes toward technological development.

The first question produced the findings shown in Table 9.

Table 9. Attitudes toward technological development

Question: If new labour-saving machines and equipment were introduced in your plant would you:

Actively support such changes?	74 %
Approve of such changes without giving them active support?	20 %
Not approve of such changes without resisting them actively?	3 %
Actively resist such changes	-
(lack of responses)	3 %

The question Are you afraid you might lose your own job because of automation? produced 15 % of affirmative answers. This, however, expressed not so much fear as actual anticipation. At the same time, 91 % of the respondents stated that if new machines replaced their work they would "expect" or "rather expect" to get a new acceptable job. There were 5 % negative answers and about 3 % "don't know".

The third question asked the respondents to evaluate the effect of automation upon seven spheres. Table 11 contains the findings.
Do NA workers share the same views on the effects of automation with A workers? Numerous studies show that the closer one gets to automation the less favourably one sees its effects and the less common become favourable opinions.(Table 10)

Workers not personally involved in higher mechanisation and automation (NA) have more favourable views on the effect of automation on reducing physical effort in work as well as on mental nervous strain.

Table 10. Effect of automation on some work features

	Increases	No Change	Decreases	Don't know
1. Job security	17	28	43	12
2. Workers' participation in making decisions	16	33	18	33
3. Young workers' prospects	53	19	13	15
4. Physical strain	6	11	79	4
5. Mental and nervous fatique	33	14	43	10
6. Satisfaction with work	61	20	10	9
7. Importance of trade unions to workers	22	51	2	25

Table 11. Workers' views on effects of automation

	rho	percentage of favourable views among NA	A
Workers' participation in management will increase	0.04	21	25
Prospects for young workers	0.09	54	64
Physical strain	0.31	96	74
Mental and nervous strain	0.17	62	42
General satisfaction	0.04	67	69
Greater importance of trade unions	0.02	29	28

The favourable opinions on the effects of automation held by NA workers do not mean, however, that the A workers would less gladly accept the introduction of new labour-saving machines and equipment. 70 % of NA workers and 90 % of A workers responded: "I would actively support such changes" (rho = 0.234).

6.1. Reasons Behind Workers' Expectations on Automation

It seems to be illustrative to take into account the reasons given by the workers at the same time when studying the distribution of answers presented in Table 11. These reasons[8] were given in a parallel study conducted among about 150 workers.

Job Security:

(a) Automation reduces job security.
The reasons given deal with the increase of workers' mobility due to the transfer of workers replaced by new machinery and due to new demands and sometimes a decrease of wages which stimulates workers to move.
"Norms are on the increase as well as requirements and not everybody likes it; stability is diminishing."
None of the reasons given suggested an unemployment scare or anxiety about finding a job.

(b) Automation does not affect job security
"There is always a shortage of workers."
"There is always a shortage of workers and some old machines will stay, nevertheless, so if one wishes to, one will be able to work them."

(c) Automation increases job security
"A worker when skilled is priceless."
"When somebody gets a good job with a good machine, he tries to keep it."
Four examples of controversial opinions about the issue: will workers have bigger or smaller shares in making decisions about matters concerning their own workplaces under new technological conditions?
"Experience and training help make good decisions."
"Technology is definite and the worker must closely follow the technology."
"There is always a predetermined way of working."
"It doesn't depend on the machine - it doesn't matter whether it is new or not."

Here are some examples of responses that illustrate workers' views on the effect of automation on the chances of young workers, on mental and nervous strain, and on satisfaction with work.

Chance for Young Workers:

(a) Automation improves the chances of young workers:
"They have to know more, they will earn more and will hold better positions in the future."
"They are less tired physically, know more, will stay healthy longer."

(b) Automation reduces the chances of young workers:
"Work with automatic machines is always the same, it does not make the worker raise his abilities and thus reduces his promotion prospects."

Mental and Nervous Strain:

The effect of automation on mental and nervous strain:
"Machines are more complicated so the mental strain is bigger."
"You have to pay attention all the time, everything has to be in keeping with technology, a lot of lights, gadgets."
"The machines are very complicated, when something goes wrong it is difficult to repair them."
"The machines often fail and people get nervous."
The responsibility for an expensive machine is greater."
"A high rate of work, the work is monotonous and tiring."

Standing and watching makes you tired."
"If the machine works well you can take it easy."
"You do not have to think, everything has been fixed in advance."

Job Satisfaction:

(a) It increases satisfaction:
"The work is lighter, less dirty, easier."
"The machines are nicer and easier to handle."
"The work is interesting, there are few such machines in the plant."
"You do better work because machines are more precise and you can have a chat with your friend, you can sit down and your legs don't get swollen as you stand."
"You're more satisfied because the quality and production output increase."
"If one works such a machine it proves one is a good specialist."

(b) Reduces satisfaction:
"After a few years work with an automatic machine, you get sick of it because of monotony."
"If you are unskilled, you don't know what kind of job you'll get later."
"You have to watch all the time and it makes you tired."

(c) Automation has no effect on satisfaction with work:
"You're less tired physically, but more mentally and it's one and the same thing."
"The work is lighter, better, but the pay is lower so satisfaction is the same."

6.2. Correlates of Attitudes Toward Automation

One of the questions which appears interesting reads: What is the relation between views on the effect of automation on physical, mental, and nervous strain and on opinion about the effect of automation on satisfaction with work? In other words: Does the positive effect of automation in such aspects as satisfaction with work correlate with positive views on other automation effects?

Figure 7 contains an answer to this question.

The horizontal axis carries three categories of respondents:
- The first (marked with number 1) embraces those who expect that automation reduces satisfaction with work; in our study the group constituted 9 % of the sample.
- To the second group (2) belong those respondents who stated that automation does not affect satisfaction with work (19 % of all respondents).
- Group number three (3), the most numerous (61%) claimed that automation increased satisfaction with work.

The vertical axis shows the percentage of respondents for each of these three categories who answered questions concerning the effect of automation on four aspects of work.

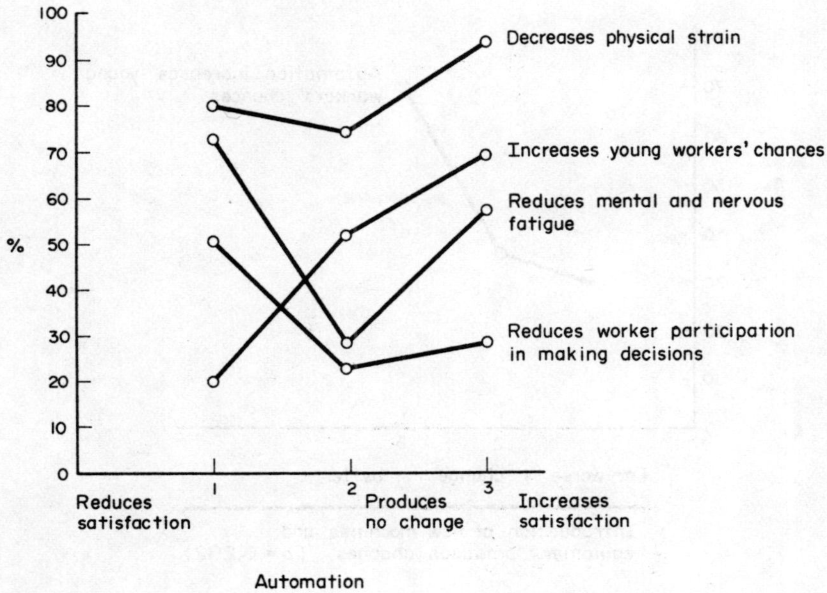

Fig. 7 Relationship between anticipated satisfaction with work resulting from automation and other automation effects.

The strongest positive correlation, which is reflected also in the course of the curve, occurs between the increase in satisfaction and the increase in prospects of young workers. There is no relation whatsoever between views on the effect of automation on mental and nervous strain and evaluation of the effects of automation on satisfaction with work (group 3); and those who anticipate a decline in satisfaction with work (group 1) are also of the opinion that automation will cause a reduction of mental and nervous strain. In general, the anticipated increase in satisfaction with work is connected with the two effects mentioned above.

To what extent do the effects of automation anticipated by workers connect with the respondents' experience?

To check the relations between experience and anticipated effects, let us have a look at responses to the question: How did the introduction of new machines and equipment affect the situation of workers in this plant? Generally speaking the change has been for better, for worse, or no change at all.

It follows from the analysis that there was only one stronger relation between responses to this question and the anticipated prospects for young workers (rho = 0.292; Fig. 8). A slight relation occurred between experience and physical strain (rho = 0.105; Fig. 9). There was also a slight relation between experience and the effect of automation on satisfaction with work (rho = 0.096; Fig. 10).

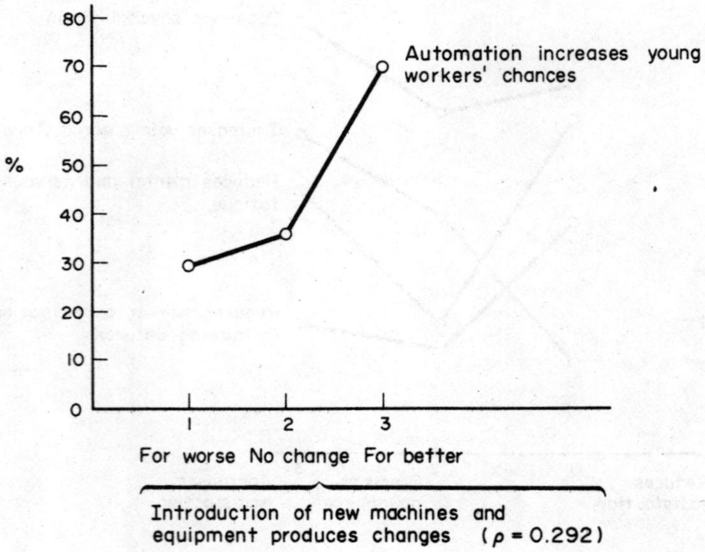

Fig. 8 Experience and anticipated positive effects of automation on young workers' chances

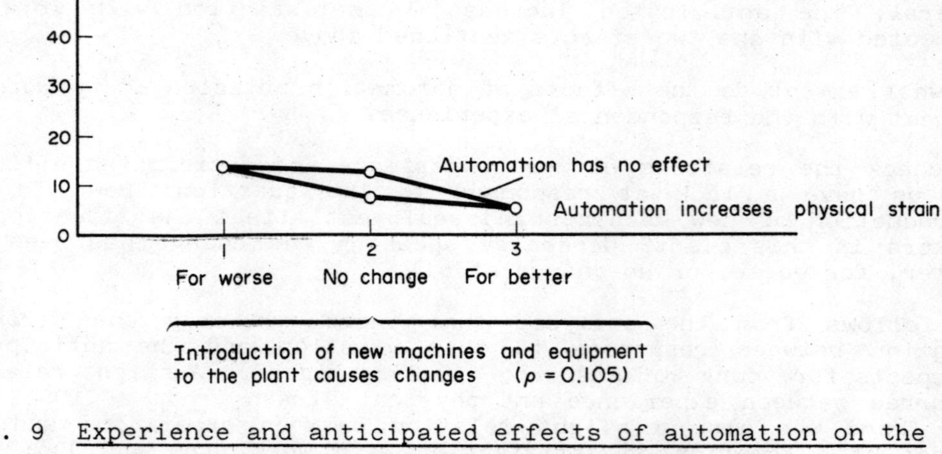

Fig. 9 Experience and anticipated effects of automation on the level of physical strain in workers' work.

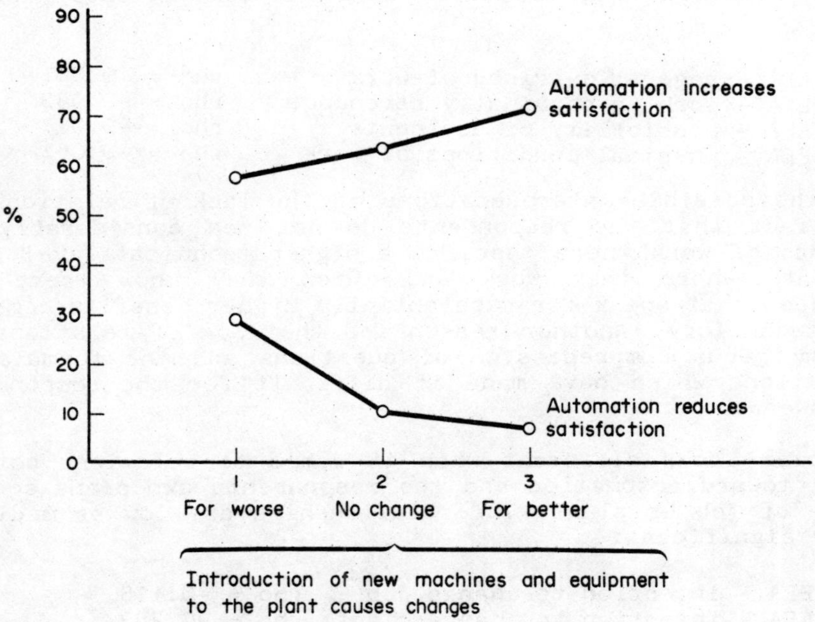

Fig. 10 Experience and anticipated effects of automation on satisfaction with work.

Thus far, only individual relations between experience, as indicated by answers to this question, and anticipated single effects of automation have been considered. Now the findings of an analysis carried out between individual experiences and expected effects of automation (EEA) will be presented.

The EEA index was constructed on the basis of ratings which the respondents gave to five aspects shown in Table 10. The respondents who gave 4 or 5 positive ratings were put into the group "extremely favourable" to automation. The respondents who gave 2-3 unfavourable ratings (e.g. automation increases physical strain, reduces prospects for young workers) were placed in the "least favourable group." The remaining were included in group 3, an intermediate one. The size of these three groups is as follows: "most favourable" - 29 %; "intermediate" - 50 %, "least favourable" - 21 %.

Undertaking the analysis of the material it was possible to assume for instance that:

(a) the worse the situation of the respondent at his workplace (e.g., concerning physical strain), the more favourable he will be to automation effects, anticipating that automation will improve his work situation;
(b) the respondents who are eager to improve their situation by changing their job or plant will be less favourable to automation effects.

The first assumption was not confirmed: the relations between EEA and selected work aspects proved weak and statistically insignificant, e.g.:

EEA - general overload of work	rho = 0.001
EEA - work is physically strenuous	rho = -0.033
EEA - possibility of accidents	rho = -0.017
EEA - physical conditions of work	rho = -0.061

One of the possible interpretations of the lack of relationship may be the fact that the respondents do not see considerably better conditions of work where there is a higher technical level, that in the plants where they work and which they know there are no differences that speak for substantially higher benefits coming from higher technology. Another reason for the lack of relationship may be an ambiguous comprehension of questions and the formulation of the questions which have made it difficult for the respondents to give answers.

The problem looks different when it comes to relations between an attitude toward automation and the respondents' own plans concerning a change of job or plant. The relationships are low or medium, but they are significant:

EEA - intention to change job	rho = -0.115
EEA - intention to change plant	rho = -0.283

The following percentage of those who intend to change job and plant appeared in three categories of respondents according to their attitude toward automation:

	Percentage of respondents who intend to change	
	The job (%)	The plant (%)
Among:		
those least favourable to automation	52	41
moderately favourable to automation	50	12
most favourable to automation	37	9

Much stronger relations have been found between the global evaluation of EEA and the workers' views on the management system. The sequence of these relations according to their strength is as follows. (See also Figs. 11 and 12).

Global evaluation of automation effects	Is the management of this company willing to consider the workers' ideas and suggestions?	0.302
	Does the company try to improve working conditions?	0.228
	Does the company take an interest in the personal welfare of the workers? (e.g. pay, housing, individual problems, etc.)	0.241
	Is the work process sensibly organised at this company?	0.196
	Are the equipment and resources with which the workers do their work at your plant efficient and well maintained?	0.185

More general conclusions from these findings may be made as follows: a favourable opinion about the effects of automation depends, to a large extent, on the system of management and in particular on the management being open to workers' ideas and suggestions and on management's interest in bettering working conditions and workers' standard of living. A belief in management acting for the workers' benefit adds optimistic colours to the effects of changes which are likely to be made in the plant or in this case by automation.

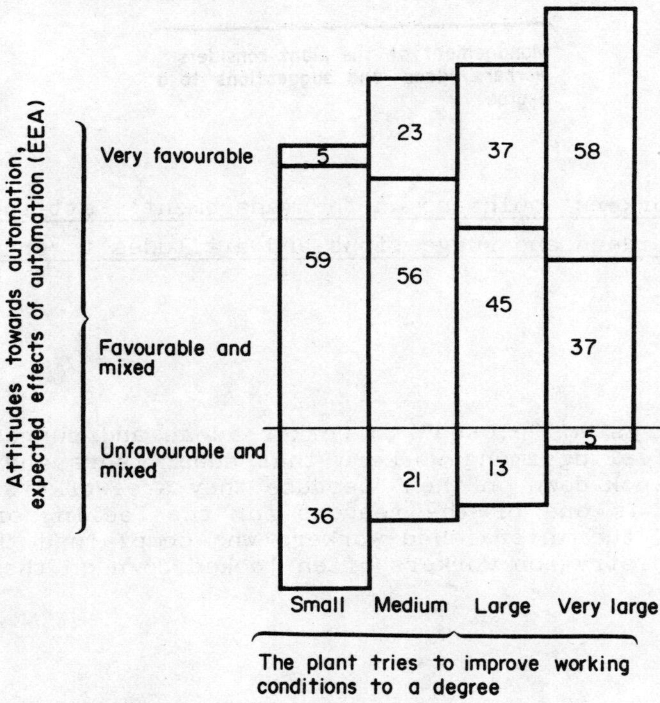

Fig. 11 The plant's interest in working conditions according to the workers' views and their attitudes towards automation.

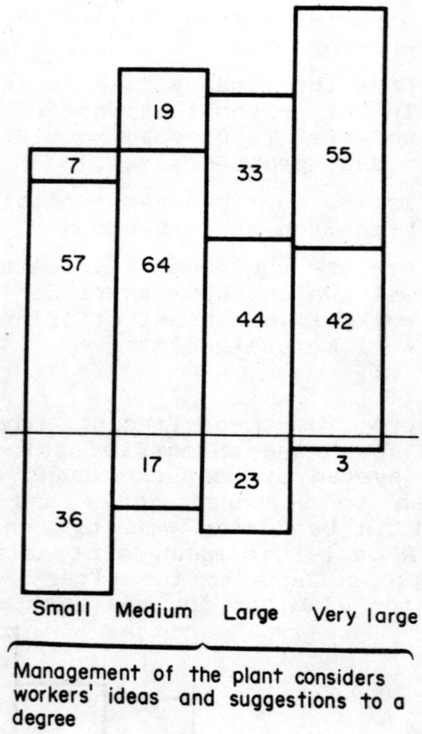

Fig. 12 The workers' opinions on the management's attitude towards their ideas and suggestions and attitudes towards automa - tion.

A management system "closed" to worker ideas and suggestions tends to create a feeling among workers that supervisors and administration workers look down on them "because they are workers".This style of management is one of the reasons for the feeling of social gap among 15 % of the interviewed workers who complained that supervisors and administration workers often looked down on them (Fig. 13).

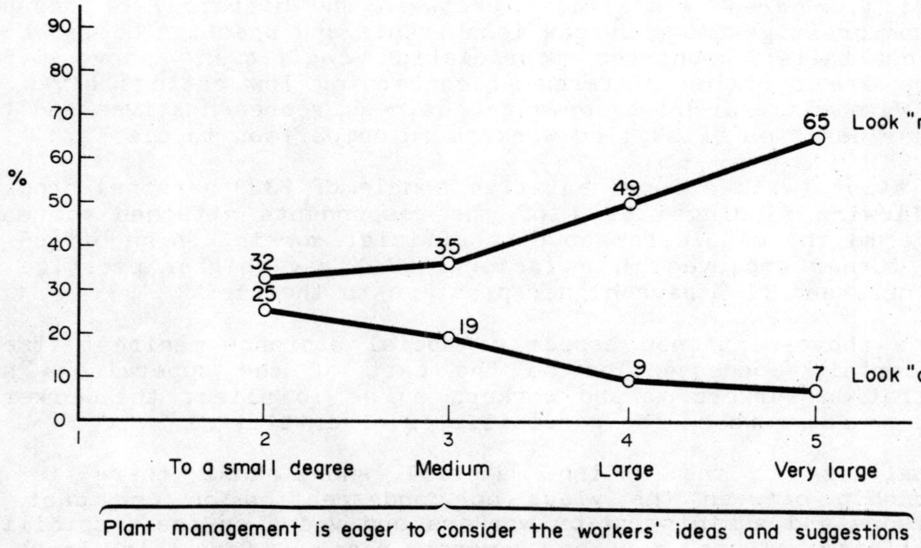

Fig. 13 The workers' opinions on the plant management's attitude towards their ideas and suggestions and frequency of a notion that supervisors and administration workers look down on the worker.

However, other perhaps more important reasons for such opinions have to be mentioned: some cultural traditions of Polish society such as large social distances.

These large pre-war social distances have considerably lessened under new social, economic, and political conditions. Nevertheless, some traces of these distances and biases are still to be seen. To illustrate and present our findings against the background of more general social processes, let me give several synoptic sociological findings.

In the studies covering representatives of Warsaw (1958) and Lodz (1961),[9] I found that:

- during the last 30 years essential changes in the position and prestige of occupations have taken place in Poland;
- apart from great similarities between the hierarchy of occupational prestige among Warsaw inhabitants and prestige hierarchies in capitalist countries (correlation coefficient above 0.8), there are striking differences concerning low ratings given in the Warsaw hierarchy to private business representatives and the higher position of skilled workers in comparison to clerks.

A 1973 study (with a representative sample of 2333 persons) brought the following findings: 52 % of the respondents attached the same prestige to the middle-rank public official working in an office as to the turner employed in a factory, 37 % gave higher prestige to the turner, and 11 % gave higher prestige to the clerk.[10]

Does the above-mentioned aspect of social distance manifest itself in a certain condescension on the part of the supervisors and adminstration workers toward workers so as to affect the workers' attitudes and views on the possibilities of criticism?

The analysis of the Polish material shows that there is no relationship between the views on condescension on the part of supervisors and administration workers and views on the possibility of criticism. There may be several reasons for this lack of relationship. The argument that condescension limits criticism can be countered with a statement that under some circumstances this attitude stirs and increases criticism. Polish data concerning a question on the possibilities of public criticism suggest the need to include some specific cultural features which might have affected the findings.

Public criticism is not commonly treated in Poland as the only, universal, and best way to act efficiently. I found in the 1969 study, which covered more than 2400 workers, that according to them some problems should mainly be dealt with in small groups - and this is not considered as public criticism. Others should be tackled directly ("I would point it out to him personally"), other problems should be turned over to the superiors, and others to a public meeting.[11]

For instance, the following view: "I would have the superior learn about it" was shared by 69 % of persons as a response to a theft affecting the plant. 13 % said "I would criticise him in public"; 10 % "I would point it out to him personally".

A tendency to criticise in public was the most common reaction to favouritism (47 %) and careless work (32 %). This does not mean that other forms of active disapproval and conduct were of little importance.

At the same time these studies confirmed a great involvement, responsibility, and attitudes of active disapproval among workers (among more than 90%).

It seems that the importance which Polish workers attach to criticism in public as an effective, moral means to create changes considerably affected responses to the question on the possibilities of public criticism in the plant. Taking into consideration favourable ratings given to co-workers, one can easily grasp the findings: about 63 % of the respondents claim that workers think they may criticise other workers in public. "May criticise" has a special ethical connotation in Polish. Some respondents answered the question:"they may, but they shouldn't," by which they mean not the danger they incur, but a reluctance to use the way which is sometimes spectacular and approved of, but unpleasant for the criticised.

7. FINAL REMARKS

We have collected the views of 109 workers holding NA jobs and those of 97 workers in A positions. The detailed opinions of both categories of workers on individual aspects of work are very similar and the rare differences which occur are in most cases in favour of NA workers.

Does this mean that a higher technological level does not change anything for the better? What is more, perhaps it deteriorates the workers' situation? Does the spectre of automation which is frequently talked about in the West also loom ahead in socialist countries? The answer to these questions based on our study seems to be negative.

Global satisfaction with work is the same among A and NA workers. Readiness to change the job or plant is equally common for A and NA workers. In most cases the opinions of both categories of workers are alike and the differences are insignificant. Both categories of workers evaluate the effects of automation in more positive than negative categories, and both groups of workers are definitely in favour of new technology and are ready to back its introduction actively. They see automation and new technology as something unavoidable and beneficial for the society, plants, and themselves. There is no conflict between the machine and man, between management who want to introduce new technology and workers who fight it. There was not a single opinion in any interview that new technology is, for instance, or might be, a means to exercise control over workers. The prevailing acceptance of new technology by workers proves that they find it beneficial and useful.

Let us try to explain the reasons for the similarities between NA and A worker's opinions on many aspects of work:

(a) The same social and political system makes general conditions that affect the situation of both categories of respondents uniformly; it also standardises the management system in the plant and working conditions.
(b) The investigated units (NA and A departments) are located in the same workplaces with a relatively short history; these are rather or really modern plants. Our study was carried out in three workplaces of the machine industry and in each of them in A and NA units. The exclusion of plants with uniform technology of a conventional kind, at a low or intermediate level of technology, gives, a priori, limited possibilities of getting fuller answers to the central question of the project.

(c) The plant management is interested in the situation where there are no substantial differences in working conditions in order to avoid just and undesirable complaints from workers, tensions, and such harmful phenomena as, for example, excessive labour turnover or excessive internal mobility.

(d) The workplaces with higher technology (A) are partly automated; the A workers still perform strenuous operations.

These reasons probably contribute to the fact that NA and A workers' views are considerably similar. However, there are some differences, and it is of interest to explain these differences, particularly where A workers are less favourable. It seems that the solution lies not only and mainly in the work at NA and A workplaces or the age of the respondents which is similar in the examined NA and A units, more important might be the level of education of respondents (A workers have higher education), level of pay (NA workers are better paid), evaluation of supervisors (NA workers give more favourable opinions).
General conclusions of our study can be reduced to an utterance made by one of the above-quoted workers: "It does not depend on the machine - it does not matter whether it is new or not." The idea in this statement is that non-technological factors play a more important role in the creation of a better quality of life and work. This, however, does not mean that technology plays a small role in this process. This has been confirmed by a widespread approval of new technology and a belief that it is a source of many new benefits.

* * * * * * * * * *

The studies on automation were started not only for theoretical purposes. The research carried out in three Polish workplaces enjoyed, from the very beginning, a real interest of the Ministry of Machine Industry, the management of the examined workplaces, and the Central Council of Trade Unions as well as trade union organisations of these workplaces.

Discussions have been carried out over the initial findings with the participation of the management and trade union organisation of the workplaces as well as administration workers of the machine industry and on the decisions concerning the means and ways of better use of new technology. The main emphasis was placed on some aspects of unsatisfactory working conditions, e.g. noise level, strenuousness of work, some shortcomings in the system of wages and promotion (they are now being changed), care for the rhythmical pace of production, greater possibilities of worker participation in making decisions.

A more comprehensive programme of action and use of richer and more efficient variety of means belong to the future. They require a deeper analysis of the collected material, additional studies, considerations and experiments. Despite the great complexity of these studies, nevertheless, research is easier to undertake than beneficial and efficient action. We would be happy if an international co-operation in the field of Automation and Industrial Workers also contributed to new publications that would be directly beneficial to people of the coming automation epoch.

NOTES

1 I use the term "automated workplaces" arbitrarily: in reality these workplaces should fall within partly automated, not entirely automated or highly automated.

2 Spearman's rho is a correlation between ranks over individuals. That is, in effect, an individual is assigned a rank on each of the two variables and then the ranks are correlated using a Pearson product moment r. Spearman's rho ranges from minus one to plus one. If considered data are ordinarily and not intervally scaled it is better to use ranks than original numbers.

3 R. Blauner, Alienation and Freedom. The Factory Worker and His Industry, University of Chicago Press, 1964, p. 18.

4 I have not presented the values of seven other components because they require some explanations for linguistic reasons going beyond the scope of this paper.

5 The reservation has to be made that these results are based on answers of workers with at least one year's service.

6 See, for example, studies by Pierre Naville.

7 A. Sarapata, Attitudes Toward New Technology, (in Polish), Warsaw, 1959.

8 This study was conducted among 300 skilled workers employed in three workplaces (machine industry) in order to check some of the questions of the workers' questionnaire, to gather material for verification of some theoretical-statistical models of job satisfaction, and to gather responses to twenty-four open-ended questions dealing with workers' opinions on the introduction of new technology.

9 A. Sarapata, Social Stratification and Mobility in Poland (in Polish), Warsaw, 1965.

10 A. Sarapata, Occupational Prestige Hierarchy Studies in Poland, paper delivered for the VIII World Congress of Sociology, Toronto, 1974. Similar data are to be found in research conducted by W. Wesolowski, Z. Slomczynski, and others.

11 R. Dudek and A. Sarapata, Occupational Ethics (in Polish), Warsaw, 1972.

5. Soviet Workers and Automation of the Production Process

V. I. USSENIN, V. V. KREVNEVICH, V. I. MASLOV,
M. N. NOTCHEVNIK, D. P. PODKOSOV,
V. V. KOLBANOVSKY, G. M. DENISOVSKY and
E. G. SAROVSKY

INTRODUCTION

Investigation of the effects of automation upon the situation of industrial workers, on job content and working conditions, and upon workers' attitudes toward work and life is at present of great theoretical and practical significance. The massive introduction of automated equipment and other advanced technology involves not only technological and organisational but also major social problems. Workers themselves, plant managers, and leaders of worker organisations, as well as society as a whole, need to know the trends of ongoing social processes in order to guide them, encourage positive development, and prevent negative results.

The international research study "Automation and Industrial Workers" was undertaken in the Soviet Union at four major automotive plants. In accordance with the research programme, the National Group prepared a report "Social and economic system of the Soviet Union and its influence on technological changes and their social consequences"; the appropriate information was obtained from plant managers and leaders of worker organisations. At the plants, ten production units were studied (5 automated and 5 non-automated), 187 jobs were evaluated, and 385 workers interviewed according to the standard workers questionnaire.

The distribution of respondents according to types of jobs and units (A = automated; NA = non-automated) is as follows:

Job Type Groups in Soviet Sample

	Operator	Set-up	Operator set-up	Maintenance	Foremen	Controllers	Total
A	69	46	43	22	4	-	184
NA	156	17	14	5	6	3	201

The study has confirmed the conclusions of a number of earlier studies in the Soviet Union and elsewhere to the effect that the very level of technology substantially affects many factors of man's work activity.

At the same time, the study has revealed a substantial similarity in the responses of workers from different types of production, jobs, and plants with regard to a considerable number of factors characterising their position at a plant, behaviour, and attitudes. This has led to the conclusion that the evaluation by workers of the above factors is determined by general social and economic conditions of a specific country rather than by the level of technology.

Finally, the study has shown that the social consequences of automation cannot be correctly understood if divorced from the workers' influence on the processes of technological change. Workers are not just objects affected by technology, a passive element which only feels the impact of positive or negative results of technological innovations. They themselves seek to determine the course of progress of science and technology and its social consequences. Data obtained at Soviet plants are evidence of the fact that Soviet industrial workers really perform such functions having become equal participants in making decisions on matters of production and social development.

The implementation of the international programme has made it possible to obtain at Soviet plants a wealth of information which can be analysed in various aspects and provide substantiation for many concrete conclusions. The most general conclusions of the study can be summed up in the following three main proposals:

(1) Soviet workers exert active influence on the processes of automation and their social consequences.

(2) Introduction of automation has a positive effect on job content, working conditions, education, and skill level of workers and their general satisfaction with their work.

(3) There are no substantial differences between workers of different jobs and plants or between workers of automated and nonautomated types of production, on a wide range of matters pertaining to the workers' position at a plant, social guarantees (including employment security), perception of social opportunities, the nature of relationship with co-workers (including managerial staff), estimation of the role of worker, organisations, or the attitude toward technological innovations and social activity. The general social and economic conditions of the country concerned, rather than the level of technology, are of dominant significance for all such questions.

This paper deals with these three main conclusions of the study.

1. THE WORKERS' INFLUENCE ON THE PROCESS OF TECHNOLOGICAL CHANGE AND ITS SOCIAL CONSEQUENCES

The international research was focused on studying the social consequences of automation for industrial workers.

In investigating this problem, strictly in accordance with the common international Research Design, the researchers constantly found that the social consequences of automation alone cannot be understood if divorced from the influence of workers and their organisations on the policy of technological innovations. This was borne out by information obtained from management as well as from leaders of party, union, and other organisations, and was confirmed by the data from interviews of workers.

1.1. Influence on Decision Making

The results of the study show that active influence by workers on decision making promotes successful operation of the plant, makes it possible to combine correctly the interests of the plant, worker collectives, and individual workers, and offers extensive opportunities for developing the workers' creative activity outside the immediate work functions. It is very significant that all interviewed managers were unanimous in the opinion that worker participation in decision making makes it easier for management to fulfil its tasks.

The workers' basic attitude toward this question is revealed in the fact that no Soviet worker interviewed doubted the need for worker participation in industrial management, or declared the lack of a personal wish to participate in decision making. 85 % of those interviewed were convinced that workers should have the opportunity to participate in making decisions on all matters affecting the plant with 80 % of workers expressing their own willingness to participate in such activities. The sphere of interests of the rest of the workers is connected with their immediate work activity (Table 1).

Table 1. <u>The workers' desire and willingness to participate in decision making</u>

I. Desire of workers for opportunities to participate in decision making	Code	Percentage of "yes" responses			II. Willingness of workers to participate in decision making	Code	Percentage of "yes" responses		
		Total	A	NA			Total	A	NA
1. Workers should have the opportunity to participate in making decisions on all matters affecting the plant	4	84.8	84.2	85.3	1. I am willing to take part in most decisions affecting the plant	4	79.7	79.8	79.6
2. Workers should have the opportunity to participate in making decisions only on problems affecting their work	3	11.5	14.2	9.1	2. I am willing to take part in decisions only on questions affecting my work	3	17.4	18.6	16.3
3. Workers should have the opportunity to participate in decisions only when management asks them	2	3.7	1.6	5.6	3. I am willing to take part in decisions only when management asks me	2	2.9	1.6	4.1
4. Workers need not have the opportunity to participate in decision-making	1	0.0	0.0	0.0	4. I am not willing to take part in decision-making	1	0.0.	0.0	0.0
Mean score		3.8	3.8	3.8	Mean score		3.8	3.8	3.8

The existing practice convinces workers that they have real opportunities to participate actively in decision making on all matters affecting plant activity. The workers were given a list of twenty questions and to each of them between 77 % and 97 % of workers declared that they had possibilities to participate directly or through their representatives in making decisions.[1] The workers' replies to that group of questions have statistically meaningful and logical correlations with answers to the questions which characterise the possiblitiy of criticising in public other workers of the plant, including representatives of management in their presence ($r = 0.34$).

It is noteworthy that the real possibility of participating in decisions about the development of the production plant is acknowledged by 90 % of workers interviewed, and 94 % acknowledged a real possibility to participate in decisions about the introduction of new machinery and equipment. It is those decisions that cover the most important aspects of automation and predetermine its social consequences. Undoubtedly, it is very important for workers to have possibilities to participate in decision making on social, personal, and pay matters which arise in connection with technological innovations. But to take into account all the interests of the workers and to ensure solutions of all the complex questions and problems is possible in practice only at the planning stage when prospects, directions, date, and terms of introduction of new technology are determined (Table 2).

Table 2. Is it possible for the workers of the plant to participate directly or through their representatives in decisions on the following questions?

	Percentage of total number of respondents		
	Yes	No	Don't know
1. Development of production plan	89.6	1.8	8.6
2. Introduction of new machinery and equipment	94.0	1.8	4.2
3. Work organisation and working conditions	97.1	1.6	1.3
4. Personnel planning (planning of future force demands)	77.1	6.5	16.4
5. Determination of basis and methods of wage payment	77.0	5.5	17.5
6. Worker job classification and pay scale	82.2	6.3	11.5
7. Determination of bonuses and other incentives	95.3	2.6	2.1
8. Hiring of workers	76.2	7.6	16.2
9. Promotion of workers	90.1	4.7	5.2
10. Selection of workers for training courses	95.3	2.1	2.6
11. Transfer to other work places	90.3	5.0	4.7
12. Allocation of overtime work	88.7	6.3	5.0
13. Disciplinary measures	96.9	2.3	0.8
14. Lay-off and discharge of workers	95.9	2.3	1.8
15. Cut-back of production and closing down the plant	27.5	7.3	65.2
16. Introduction of rationalisation proposals and inventions made by workers	93.5	1.8	4.7
17. Development of construction plan of housing and cultural and welfare facilities	88.0	2.9	9.1
18. Providing living accomodations in houses belonging to the State, co-operatives, and public institutions	97.9	0.5	1.6
19. Granting social insurance benefits	95.1	1.0	3.9
20. Providing accomodation in rest-houses and medical treatment in sanatoria	97.2	1.0	1.8

1.2. Influence on Planning Process

The workers' influence on planning, in effect, predetermines subsequent decisions related to the implementation of the approved plans. Of primary importance in this connection is the fact that the participation of the workers themselves in the elaboration of these plans is their main source of information about long-term plans for technological innovations. Such answers were given at all investigated plants by management and union representatives. The correlation analysis has revealed substantial statistical links between the workers' estimation of the possibilities to participate in the planning of technological innovations and estimations of general results of those innovations for workers themselves ($r = 0.43$).

At present, preparation of a special plan of social development (along with economic plans and plans for the introduction of new technology) is a specific feature of planning at Soviet plants. At the time of this study, such plans were operational at three out of the four investigated plants. These plants are drawn up by representatives of worker organisations and management, aided by scientists. The plans of social development supplement the plant collective agreement.

These plans sum up main production targets, forthcoming technological changes, and mechanisation and automation measures. On the basis of the targets, the plans envisage very detailed and concrete measures in establishing work groups, improving the workers' skills and occupational structure, changing conditions of pay and methods of wage payment, improving social and cultural services, workers, housing, humanisation of work, etc. Through such planning, workers can see clear-cut prospects of social development of their collective in the forthcoming (usually 5-year) period.

Information obtained from management and unions gives a graphic picture of the practice of decision making at Soviet plants (Table 3). On the majority of issues, the management takes decisions either after preliminary discussions with the workers' representatives or only with their consent.

Only hiring of new workers and taking disciplinary measures can be decided by the management unilaterally. In practice, however, union influence on such matters is also considerable.

At three out of the four investigated plants, the management takes disciplinary measures after preliminary discussions with the workers' representatives. The same procedure is used at one plant in hiring new workers (according to the information of union representatives). Workers themselves believe that their representatives can also participate in decision making on the questions of hiring of workers (76 % of respondents) and taking disciplinary measures (97 % of respondents). These replies indicate not only the general perception of the workers' role at the plant, but also the realisation of the workers' concrete possibilities to influence appropriate decisions.

At some of the investigated plants there are voluntary worker commissions attached to personnel-departments, which informally consider personnel matters, including hiring of new workers. At the same time, disciplinary measures, even when imposed unilaterally by the management, can be appealed by workers to the commission on work disputes, and subsequently to the union committee whose decision would in principle be final.

1.3. Institutions for Workers Participation

The system of industrial democracy of Soviet plants enables the working people to influence decision making both directly and through their elected representatives. The Communist Party of the Soviet Union (CPSU) and the Trade Union and Young Communist League (Komsomol) organisations enjoy greatest authority at the plants.

At the investigated production units, CPSU membership stood at 10-33% of the total number of industrial and office workers, with the majority of party members being industrial workers. Their special influence in the collective is determined by the fact that the party organisations are comprised of the most conscientious and active members of the work collective. These are advanced workers whose work and social activity are a model for the rest of the working people. Non-party workers are invited to participate in the activities of the party organisation. Fifteen to forty per cent of non-party workers of the investigated shops took part in party meetings along with party members. Non-party workers participate in many committees set up by party organisations.

Party organisations have the right of control over the operations of management, and thus in pratice they have possibilities to take an active part in decision making on most important directions of the plant's activities.

Trade Unions are the biggest mass organisation of workers. At the investigated plants, union membership ranges between 98% and 100% of workers, and at the investigated units union membership was 100%. The law gives most detailed powers to trade union organisations and their committees to participate in decision making on production, economic, social, and personnel matters affecting the entire plant, the workers'collective, and individual workers.

Under union committees there are standing production conferences made up of the workers'representatives, representatives of their organisations, and the management. Conference decisions on production matters are carried out by the management.

Komsomol organisations, which represent the interests of young people, participate in decision making on a most extensive range of issues.

Table 4, compiled on a basis of replies to the workers questionnaire, gives an idea of the ways used by Soviet workers to influence managerial decisions.

Table 3. Ways of making decisions on main matters of plant production activity[a]

Matters	Administration makes decisions: Unilaterally	After preliminary discussion with the workers' representatives	Only with th consent of the workers' representati
1. Development of production plan		I, II, III, IV / 2	1, 2, 3
2. Introduction of new machinery and equipment		III	I, II, IV / 1,2,3,4
3. Work organisation and working conditions			I, II, III, / 1, 2, 3, 4
4. Personnel planning (planning of future work force demands)		I, III, IV	II
5. Determination of basis and methods of wage payment			I, II, III, / 1, 2, 3, 4
6. Worker job classification and pay scale			I, II, III, / 1, 2, 3, 4
7. Determination of bonuses and other incentives		II, III	I, IV / 1,2,3,4
8. Hiring of workers	I, II, III, IV / 2, 3, 4	1	
9. Promotion of workers		IV	I, II, III, / 1, 2, 3, 4
10. Selection of workers for training courses		I, II, IV / 1	III / 2,3,4
11. Transfer to other work places		I, IV / 1	II, III / 2, 3, 4
12. Allocation of overtime work			I, II, III, / 1, 2, 3, 4
13. Disciplinary measures	III / 3	I, II, IV / 1, 2, 4	
14. Lay-off and discharge of workers			I, II, III, / 1, 2, 3, 4
15. Cut-back of production and closing down of plant		(Such cases have not occurred)	
16. Hiring and dismissal of administrative personnel		I, II / 1,2,3,4	III, IV

[a] Data representing plant administration (I-IV) and trade union representatives (1-4).

Table 4. Ways Soviet workers use to influence managerial decisions

Here are some ways that workers in different countries use to influence managerial decisions.

	In which activities did you personally participate in the last 2 years ?		
	Participated	Did not participate	No response*
1. Participation in the meetings of worker organisations (e.g. meetings, trade union meetings, meetings of the plant Party	98	2	-
2. Participation in the work of different committees at the plant, (e.g. trade union, Party, Komsomol, committees, and commissions)	57	43	-
3. Expressing your ideas to your immediate supervisor or management	89	10	1.0
4. Expressing your ideas to your representatives of Party, trade union, and Komsomol organisations	70	28	2.0
5. Contacting representatives of public organisations outside the plant (e.g. Party, trade union, and Komsomol organisations)	19	79	2.0
6. Contacting the state (governmental) bodies (Soviet of Deputies of Working People, Ministries,and Departments)	7	91	2.0
7. Arousing public attention (e.g. organising demonstrations, distribution of leaflets)	0	100	0
8. Methods of collective pressure on administration (e.g. refusal of overtime work, slowdown, work to rule, strikes)	0	100	0
9. Sending letters to plant, local, and central newspapers	15	84	1

* (in % of the total number of respondents)

Participation in the meetings of worker organisations (general, party, union, Komsomol meetings) and in various committees and commissions should be regarded as ways of direct participation in decision making. It is the general rule that each worker organisation holds at least one meeting a month with at least 50% of the organisation membership required to constitute a meeting. Naturally, almost all workers interviewed (98%) had taken part in such meetings in the last 2 years. Not every worker participated actively in discussions, but even by voting for some proposal or other they directly influenced decision making.

A higher degree of participation involves participation in the work of various party, union, Komsomol committees and commissions, and in standing production conferences. Practically all their members personally participate in the preparation and discussions of appropriate decisions. Fifty-seven per cent of the workers interviewed were engaged in such work personally in the last 2 years.

Expressing ideas to the representatives of management and worker organisations is a way of indirectly influencing decisionmaking. These ways are extensively used in practice. In certain cases workers report defects in the work to the representatives of higher worker organisations, state authorities, and editorial boards of newspapers. To influence the attitude of management, workers do not need to resort to demonstrations, let alone strikes, because the existing decisionmaking machinery provides workers with much more effective ways of exerting their influence.

The results of the study make it possible to conclude that participation in planning and in ensuring implementation of the adopted plans is the most important way by which Soviet workers influence the process of technological change and its social consequences. The broad possibilities available in these respects are actively taken up by the working people; however, the degree of activity varies from worker to worker. The further development of worker participation in production management is a current task of worker organisations.

1.4. Soviet Research on Worker Participation

Soviet scientists devote much attention to worker participation in production management. The need for such broad and active participation is widely recognised by workers themselves, their organisations, and representatives of management. So, for example, in 1972-1974 at many enterprises of various industries in some Union Republics a sociological investigation was carried out in the course f which about 10.000 workers and 800 management representatives and leaders of worker organisations were interviewed. All the plant directors and heads of the shops admitted the necessity of worker participation inmanagement. Only 10.9 % of the interviewed gave negative answers to the question "In your opinion is it necessary for all the workers to participate in decisionmaking?"In 1973-1974 at two Moscow plants of the "Plastic" scientific and production association a sociological study was made to interview top and medium echelon leaders and also workers and employees directly participating in the activities of various committees at the plant. Ninety eight per cent of the respondents confirmed the possibility and desirability of worker participation in decisionmaking on a very broad range of questions.

Some studies emphasised that participation of workers in decision-making on technological questions not only reflects favourably on production, but also creates for workers themselves an additional sphere of application of their creative energy, which cannot always be realised in the framework of immediate duties.

The special studies on worker participation in production management carried out in the Soviet Union, analysed in much greater detail many questions that are studied only in general terms in the international project. This made it possible to call attention to the fact that at a number of plants the trade unions and workers themselves are far from making full use of the opportunities available to them. Furthermore, the effectiveness of their participation in decisionmaking varies considerably, and sometimes such participation is rather formal.

The studies of Soviet scientists do not make a special analysis of the question of whether Soviet workers may participate directly or through their representatives in making decisions on production, social, and personnel problems. Availability of these opportunities, which is guaranteed by law and the entire social system of the country, is taken into account as an indubitable reality. But scientists are interested in how this system functions in practice and how intensively unions and workers exercise their rights.

Thus, for example, it was found out at the above-mentioned "Plastic" industrial association that the workers of the organisation do not participate in making all managerial decisions on technical and economic questions. At the shop level, decisions on these questions are agreed upon with worker organisations in 46 cases out of 100, at the association level in 30 cases out of 100. Representatives of both the management and the workers'organisations explained this by a desire to concentrate on the joint adoption of major decisions in order to avoid numerous meetings and co-ordinations. The correctness of such an approach is called into question.

2. INFLUENCE OF AUTOMATION ON WORK CONTENT, WORKING CONDITIONS, AND WORKER ATTITUDES

The study of the influence of automation on work content and working conditions was conducted with the help of expert analysis of the work processs and interviews of workers.

The expert analysis of the job content of industrial workers was undertaken by the Soviet Study Group in order to identify the objective requirements upon the worker of the work functions at a given workplace. The analysis involved also the study of conditions created in the process of work that influences the development of a worker's personality as well as a critical estimation of work functions. This was done to make work both productive and conducive to maximum self-realisation.

2.1. Work Content

In accordance with the purpose of the study, the analysis of the workplace observation was undertaken with regard to thirteen composite variables.[6] The basic concept of the research is that the level of automation is the main independent variable which summarises the degree of technical progress in the industry under investigation. In studying job content it characterises the degree of transfer of the worker's functions to the machine.

All nine manual operations listed in the observation form are used at the majority of non-automated production workplaces. Manual operations take up almost all of the time at non-automated jobs with the bulk (up to 55 %) falling on loading, fixing, and unloading of material as well as on immediate control of the work process (7 % of work time). Only set-up men are engaged in more meaningful operations. Loading, fixing, and unloading take in their work only 2 % of time. On the contrary, workers on automated units, as a rule, are engaged in manual operations not more than 11 % of their work time. The operators in these units spend 5 % of work time for loading, fixing, and unloading.

A considerable share of production operations at investigated automated production units is performed without the workers'* intervention. (The mean score for the level of automation index is 5.8 vs. 2.5 in non-automated units.)

This could not but influence the changing ratio between the workers' physical and mental efforts in the process of work. The general score of all functions related to mental operations of the worker of automated production appears higher than at non-automated units, while the share of physical effort is reduced. At the same time, the general working conditions connected with equipment and comfort of physical environment improve substantially.

The level of physical strain connected with lifting, carrying, pulling, pushing, retrieval, etc., of various loads went down considerably. At the same time, the improvement of equipment, and incorporation in its design of features suggested by engineering psychology, facilitates its maintenance, operation, and repair. As a result, the share of work time involving uncomfortable body postures - crawling, climbing, sideward movements - grew smaller.* The numerical value of the resulting index of physical strain on those items at automated production is 5.2 vs. 11.3 at non-automated production.

Automation also leads to a general reduction in work intensity and accordingly lessens* worker fatigue. On the whole, the index of work intensity is lower for automated production than for non--automated (1.3 vs. 5.4). The level of physical and mental efforts connected with unfavourable situation factors decreases because of the shrinking share of work time involving these factors (see Table 5).

Table 5. Unfavourable work characteristics

Items from research instrument IV	Mean score	
	Non-automated production	Automated production
Specified workplace	3.5	1.5
Repetitive activities	3.5	2.0
Length of work cycle	2.1	4.1
Being tied to a workplace	3.8	2.3
Need to follow specific set procedures	4.0	2.3
Time pressure of situation	0.5	0.7
Attention to detail	3.1	2.7
Vigilance	2.6	1.5
Hand-ear co-ordination	0.7	0.5
Visual stress	0.5	0.3

The increasing gravity of the consequence of wrong decisions leading to breakdowns and stoppages of the technological processes (i.e. increased worker responsibility for working process) contributes to the mental strain of jobs under automated production. The numerical value of the index responsibility* for working process at automated units (0.5) is higher than at non-automated units (0.2) (see Table 6).

The growing share of mental operations leads to higher requirements of worker skills* (1.9 vs 1.0) and which necessitates:

- a higher general education (from 1.7 in non-automated to 3.3 in automated units);
- increased occupational and technical training (from 0.6 to 0.9);
- on-the-job training (from 1.6 to 2.7);
- a need for maintaining knowledge at an up-to-date level (from 0.1 to 0.8).

Technical education is not as a rule required for workers in the non-automated units, while their required general education level is on the average 7-8 years of schooling. In the automated units, workers are required to have 9-10 years of general education and in some cases higher or even special technical education. At the same time, the level of required on-the-job training rises significantly in the automated units.

Table 6. Numerical values of variables in analysis of primary data of workplace observation[a]

Variables	Numerical values of variables	
	Non-automated production	Automated production
1. Variety	0.6	1.7
2. Responsibility for working process	0.2	0.5
3. Level of automation	2.5	5.8
4. Job-required classification	1.0	1.9
5. Information processing activites	0.6	1.1
6. Job-required communication	1.4	1.7
7. Use of devices and equipment	0.5	0.6
8. Sources of job information	2.1	2.1
9. Discomfort of physical working conditions	1.9	1.5
10. Structure of job	2.6	1.4
11. Level of physical strain	11.3	5.2
12. Monotony of job	1.7	0.5
13. Intensity of job	5.4	1.3

[a] For formulae for calculation of numerical values see Appendix.

Automation, by freeing the worker from immediate material processing, reduces the degree of his direct operational attachment to the production process. In non-automated units work is almost completely structured, i.e. tolerates no substantial deviations from the established procedures. The situation is different in the automated units: here the structure of job* is 1.4 against 2.6 at non-automated units. Thus opportunities are made available for greater deviations from the prescribed operational modes, for independent decisions, for seeking ways of improving production processes and for technical creativity.

This, in its turn, increased the communication required* under automated production. The communication level index value in the investigated automated units (1.7) is higher compared with the index value in non-automated production (1.4).

Two trends which manifest themselves in the conditions of automated production contribute to this increased communication. First, a worker is less confined to his workplace (from 3.8 down to 2.3). This increases the possibilities for contact with other workers (from 1.5 to 1.7). Second, the structure of the requirements on the worker undergoes changes. In view of increased complexity of tasks and equipment, the need for production contact with other workers becomes greater (from 2.2 to 3.2). In other words, an automated line, as one single complex mechanism, requires more co-operation nd co-ordination among all those who work in the unit. In the general view of Soviet specialists, automation leads to an enriched content of worker labour, to positive qualitative shifts in the structure of the working class, and diminishes the differences between mental and physical labour which is currently evolving in Soviet society.

Other Soviet sociological studies cite data to the effect that physical labour in non-automated production consumes 66-85 % of the working time, while in automated production it is reduced in some cases to 5-11 %. In particular, these stressed the functions where mental and physical labour are merged (with set-up men on automated lines these merged functions take up 40 % of the time).[7]

2.2. Working Conditions

Automation of production has, in these cases, resulted in significant positive changes in the physical working conditions. Automatic emulsion feeding and collection of shavings precludes direct contacts with raw material, greasy and soiled parts, and equipment components. Workers in automated production plants are better protected from the impact of harmful factors in the physical environment such as high humidity, emulsion vapour, and greases. Therefore, physical conditions in automated production are, in a number of aspects, better than in non-automated production. In no single respect is impoverishment observed in the automated units (Table 7).

Table 7. Physical working conditions

Items (workplace observation	Mean score: Non-automated production	Mean score: Automated production
Air contamination	2.8	1.7
Vibration	1.2	1.2
Inadequate illumination	1.1	1.1
Dirty environment	3.3	2.5
Wetness and/or humidity	2.4	1.7
High temperature	1.0	1.0
Awkward or confining space	1.8	1.3

This decreases the risk of occupational diseases and of accidents and injuries (from 2.9 to 2.4). The total value of the resulting index of physical discomfort in automated production (1.5) is lower than in non-automated production (1.9).

Thus the analysis of job content convincingly testifies to the fact that transition from mechanisation to automation improves the content and conditions of work, increases the role of mental operations and makes the working process more varied and the worker more independent. At the same time, the level of the required skills of the worker becomes higher.

2.3. The Workers' Perception of the Changes in Work Content and Working Conditions

To evaluate the impact of automation on workers it is important to take into account not only expert opinion regarding changes in objective characteristics of work but also the opinions of workers themselves who are directly affected by automation.

The way ongoing technological change is perceived by the workers and their attitude towards this process is reflected in the workers' evaluation of different aspects of their job. By comparing ratings of workers in automated and non-automated production units we may see whether automation, from their viewpoint, makes work more meaningful, more interesting, and brings greater satisfaction to their work (Table 8).

Judging by the estimation of workers directly affected by technical progress, as can be seen from Table 8, automation improves all the basic characteristics of the work content. It increases the variety of work, makes it more independent, provides the chance to use a greater degreee of accumulated knowledge, increases the possibilities for creative work, is conducive to developing worker abilities, requires that a worker constantly enrich and expand his knowledge, and makes the work interesting for an overwhelming majority of workers.

In order to determine to what degree workers are satisfied with the above changes, it is necessary to take into account the level of worker demands since satisfaction is the relative fulfilment of needs. The difference between the rating of the importance of each factor by a worker and the rating of its presence in his work will constitute the degree of his needs fulfilment in this factor, i.e. the degree of satisfaction. In this case, the greater the difference the smaller the degree of satisfaction, and vice versa. Let us analyse the mean scores to estimate the real character and significance of each factor for the worker (Table 9).

The analysis demonstrates that non-automated production unit workers are less satisfied with their work content. The greatest difference between the level of needs and their fulfilment is noted in the need to learn new things (2.9), to develop one's own abilities (2.8), to work out new and better ways to do one's job (2.7), and to have interesting work (2.6) and variety in work (2.3).

Workers in automated units are more satisfied with all factors of job content in spite of an increased level of their needs - possibly caused by a higher level of general education and vocational training.

Thus automation of production is conducive to eliminating, or substantially decreasing, the level of dissatisfaction with job content. In automated production units the main source of dissatisfaction shifted from job content to working conditions. Though working conditions have noticeably improved, the workers' requirements are even higher.

The analysis of the workers' answers confirms that automation results in a variety of changes in working conditions (Table 10).

Automation favourably affects the character of work loads. The share of physically strenuous work decreases, as does the amount of work considered to be an overload. On the other hand, the share of mentally strenuous work tends to increase.

According to expert opinion, safety and hygienic conditions (the possibility of accidents and injuries and the physical environment) are somewhat better in automated than in non-automated units, and workers confirm the findings though to a lesser degree. Work schedules, however, are estimated as somewhat more convenient by workers in non-automated units.

The workers' desires as to working conditions in automated and non-automated production coincide in many aspects (Table 11). The majority of workers have some preference for work that is not physically strenuous, and tend to prefer mentally strenous work. Though safety and hygienic conditions in automated production units are generally high, aspirations for even better conditions are expressed by workers.

Juxtaposing worker perceptions of actually existing working conditions with their importance to them, enables one to compare levels of satisfaction of workers in automated and non-automated units. In automated units there is less difference between worker evaluations of work loads and of their importance. There is greater difference between evaluation of safety and hygienic conditions and their importance. Consequently, workers of automated units are more satisfied with work loads and less with safety and hygienic conditions than workers of nonautomated units.

Thus working conditions connected with automation are perceived as positive by workers. At the same time, workers in automated units have greater requirements regarding working conditions.

The analysis of answers by workers in workplaces with different technological levels demonstrated that, in general, workers in automated units are more satisfied with their work than workers in non-automated units. In non-automated units only 68 % of workers replied that they were "very" or "somewhat" satisfied with their work, while in automated units the same answer was given by 85 % of workers. In the latter units there were only half as many workers who were partly or completely dissatisfied with work, although in both production units this group of workers is relatively small.

Table 8. Ratings by workers of different factors characterising work content

(Percentage of total respondents in each unit)

Factors	Non-automated (NA) / Automated (A)	Very low score 1–2	Low score 3–4	Middle score 5	High score 6–7	Very high score 8–9
Variety	NA	21	30	12	25	11
	A	3	17	28	33	19
Independence	NA	4	17	17	32	31
	A	1	7	10	41	41
Responsibility	NA	1	23	15	30	31
	A	1	4	10	34	52
Chance to use your knowledge	NA	11	37	15	23	16
	A	1	4	18	45	22
Possibility of working out new and better ways to do your job	NA	24	49	13	11	3
	A	9	31	20	34	7
Need to learn new things	NA	28	37	11	14	10
	A	3	15	29	26	27
Interesting work	NA	18	27	15	25	15
	A	2	14	16	35	33
Chance to develop your own abilities	NA	17	35	19	23	9
	A	4	15	18	39	24

Table 9. The extent of the workers' satisfaction with different factors of job content
(Mean score)

Factors	Non-automated units			Automated units		
	Mimp[a]	Mext[b]	Mimp-Mext[c]	Mimp	Mext	Mimp-Mext
Variety	6.8	4.5	2.3	7.3	5.7	1.6
Independence	7.1	6.1	1.0	7.5	6.9	0.7
Responsibility	6.8	6.2	0.6	7.2	7.3	0.1
Chance to use knowledge and training	6.8	4.9	1.9	7.2	6.2	1.0
Possibility of working out new and better ways to do a job	6.4	3.7	2.7	7.1	5.0	2.1
Need to learn new things	6.9	4.0	2.9	7.6	6.0	1.6
Interesting work	7.5	4.9	2.6	8.1	6.5	1.6
Chance to develop your own abilities	7.3	4.5	2.8	7.7	6.0	1.7

[a] Mimp - mean score of importance to workers.

[b] Mext - mean score of extent to which factor is provided in a given job.

[c] Mimp-Mext - dissatisfaction value based on mean score.

Table 10. Evaluation by workers of working conditions

(Percentage of total respondents in each unit)

Factors	Automated production (A) / Non-automated production (NA)	Very low score 1–2	Low score 3–4	Middle score 5	High score 6–7	Very high score 8–9
Physically strenuous work	A	6.0	37.0	28.8	20.1	8.1
	NA	8.5	23.4	11.9	37.3	18.9
Mentally strenuous work	A	8.7	18.5	26.1	34.2	12.5
	NA	28.4	33.8	17.4	16.0	4.5
An overload of work	A	7.6	25.6	33.7	23.9	9.2
	NA	8.0	22.3	22.4	39.8	7.5
Likelihood of job-related accidents and illness	A	25.8	38.6	20.7	13.5	4.3
	NA	24.1	36.2	14.6	18.6	6.5
Good physical conditions (temperature, cleanliness, noise level, etc.)	A	3.8	14.2	31.5	33.1	17.4
	NA	3.5	25.5	18.5	35.0	17.5
Convenient hours of work	A	3.3	5.5	16.4	19.1	55.7
	NA	2.5	6.0	7.5	21.0	63.0

Table 11. The importance of working conditions to workers of automated and non-automated production

(Percentage of total respondents in each unit)

Factors	Automated production (A) / Non-automated production (NA)	The extent of importance				
		Very low score 1-2	Low score 3-4	Mean score 5	High score 6-7	Very high score 8-9
Physically non-strenuous work	A	2.7	12.0	31.0	34.3	20.0
	NA	6.0	13.5	14.0	35.5	31.0
Mentally non-strenuous work	A	32.6	28.8	21.8	10.3	6.5
	NA	36.5	29.0	17.0	10.5	7.0
Absence of an overload of work	A	2.2	10.3	29.9	31.0	26.6
	NA	1.9	11.5	13.5	29.0	30.5
Safe and healthy working conditions	A	2.7	7.6	7.1	20.1	62.5
	NA	2.0	9.5	11.6	26.2	50.8
Good physical conditions	A	0.5	5.4	7.1	20.1	66.9
	NA	1.0	10.5	10.1	27.6	50.7
Convenient hours of work	A	2.2	6.1	10.9	22.8	56.0
	NA	3.0	6.0	14.1	30.6	46.2

Table 12. Rank of factors characterising different aspects of work according to their influence on general job satisfaction

(on the basis of coefficient E2)

Rank by the extent of their influence	Non-automated units Factors	E^2	Automated units	E^2
1.	Interesting work	28.5	Interesting work	27.3
2.	Opportunity for upgrading or promotion	17.2	Mentally strenuous work	15.4
3.	Chance to develop your own abilities	14.2	Chance to develop your own abilities	14.2
4.	Chance to improve your occupational skills	14.1	Chance to use your knowledge and training	10.4
5.	Chance to use your knowledge and training	13.0	Opportunity for upgrading or promotion	9.9
6.	No overload of work	9.4	Does the supervisor give sufficient explanations and instructions?	9.5
7.	Good physical conditions	8.8	Does the supervisor see you get justly rewarded for your work?	9.4
8.	Does the supervisor listen to you when you have something important to say?	8.6	The supervisor has sufficient skills	8.3
9.	Job variety	6.7	The supervisor listens to you when you have something important to say	8.2
10.	Good pay	6.6	Help each other in getting work done	8.0
11.	Possibility of working out new and better ways to do a job	6.6	Need to learn new things	7.9
12.	Need to learn new things	5.8	Good pay	7.4
13.	Mentally strenuous work	5.6	Technical education	7.1
14.	The supervisor sees you get justly rewarded for your work	5.2	Job variety	6.7
15.	Age	4.9	Work schedule	6.3
16.	The supervisor gives sufficient explanation and instructions	4.8	Possibility of working out new and better ways to do a job	3.9
			Other factors have less significant influence or no influence at all	

Still more obvious becomes satisfaction with work at a given workplace from answers by workers to questions: Would you like to have a different job than the one you have now, and Do you actually intend to change your job during the next 12 months ?

Over 57 % of workers in non-automated production units said they would like to have a different job. In automated units the same statement was made by only 38 %. The number of workers who intend to change their jobs within the next 12 months is also less in automated units-10 % as compared to 15 %.

2.4. Factors Determining Job Satisfaction

The analysis of variance demonstrated that a level of job satisfaction, a desire to have another job, or an intention to change one's job within several months depends, in the first place, on how interesting is the given job for the worker. According to the Eta coefficients, which determine the probability of each factor's influence on the mentioned variables, "interesting work" received first rank in both types of production under investigation (Table 12).

In second place, according to its influence on the level of general satisfaction of workers in non-automated production, is "opportunity for upgrading or promotion". Upgrading or promotion ultimately ensures higher pay. However, pay is not the only aspect which determines the significance of this factor, because "good pay" by its influence on general job satisfaction occupies but the tenth place, while "upgrading or promotion" is in second place. The main reason seems to be that upgrading contributes to raising a worker's prestige. It gives him an opportunity to obtain a more interesting, more complicated, and varied job. Simultaneously, he may obtain greater independence in work. The attractiveness of upgrading and promotion results from changes in work itself, which are highly appreciated.

In the third and fourth places, according to their influence on the level of general satisfaction, are such factors as "chance to develop one's own abilities" and "chance to improve one's occupational skills" (apart from upgrading or promotion). In the fifth place is "chance to use one's knowledge and training". Work brings satisfaction only when it creates conditions, or at least provides possibilities, for developing one's abilities, improving occupational skills, and using one's knowledge and training.

In automated units, interesting work also occupies the first place as to the rank of influence on the level of general satisfaction, but demands for this factor (as well as for other factors characterising work content) are considerably higher.

In second place is the level of "mentally strenuous work". Soviet workers interpret positively the notion of "mentally strenuous work" and perceive it as indicating a desirable requirement. Therefore, the higher level of "mental strain", the greater is job satisfaction. Workers in non-automated units display similar attitudes. But for them this factor is not so important.

In the third and fourth places, according to their influence on the level of general satisfaction, are such factors as "chance to develop one's own abilities" and "chance to improve one's occupational skills" (apart from upgrading or promotion). In the fifth place is "chance to use one's knowledge and training". Work brings satisfaction only when it creates conditions, or at least provides possibilities, for developing one's abilities, improving occupational skills, and using one's knowledge and training.

In automated units, interesting work also occupies the first place as to the rank of influence on the level of general satisfaction, but demands for this factor (as well as for other factors characterising work content) are considerably higher.

In second place is the level of "mentally strenuous work". Soviet workers interpret positively the notion of "mentally strenuous work" and perceive it as indicating a desirable requirement. Therefore, the higher level of "mental strain", the greater is job satisfaction. Workers in non-automated units display similar attitudes. But for them this factor is not so important.

In third and fourth places in automated production units are "chance to improve one's abilities and "chance to use one's knowledge and traning". These factors are equally important for all workers. In fifth place is "opportunity for upgrading or promotion". For workers in automated production units, this factor is also important not so much because of higher income but because upgrading or transfer from the position of an operator to that of a set-up man ensures a more interesting job.

Thus one may assert that job content, and in particular whether it is "interesting", is the principal factor which determines the level of their satisfaction with work for workers of both types of units. Material incentives, organisation, and working conditions have less influence on general job satisfaction.

Consequently, automation of production increases the meaning of work and contributes also to greater general satisfaction. This, in turn, creates an attitude toward work as a value in itself.

This is in line with other Soviet studies.[8] A. G. Zdravomyslov gives the following factors hierarchy in terms of the extent of their influence on job satisfaction (the index indicates the weight of a given factor):[9]
"The content of the job, the creative possibilities of work are the specific factors which determine a worker's attitude to work." Such is the conclusion made by the authors of the research on Man and His Work.[10]

Job content	0.72	Work organisation	0.38
Size of wages	0.61	Management attitude to worker	0.35
Opportunity for advanced training	0.58	Extent of physical work-load	0.32

3. THE INFLUENCE OF GENERAL SOCIAL AND ECONOMIC FACTORS ON THE PROCESSES AND SOCIAL CONSEQUENCES OF AUTOMATION

The introduction of automation and other modern technologies into production is unavoidably connected not only with technical but also with organisational changes, with changes in specific tasks of worker organisations, in relations among the people at work, and in the activities of workers. Although automation is a direct impetus for these changes, the character, content, and direction of social processes connected with it depend to a decisive degree on the general social and economic conditions of the country, and on the system of relations of production.[11] This conclusion has been confirmed by the results of the research work carried out (Table 13).

Substantially differences in job content, in working conditions, and the workers' satisfaction with the work process itself discovered between automated and non-automated production units demonstrate that the level of technology exerts a direct and determining influence on these factors. On the other hand, the study of many other aspects of production and social life of worker groups indicates that technological innovations by themselves do not inevitably bring about specific social results. Rather a wide choice of solutions is permitted. In practice, the choice that is made will depend upon specific features of the social and economic system of a country.

The significance of general social and economic conditions is demonstrated by the fact that no substantial discrepancies exist between workers of automated and non-automated production units or between workers of different professions and plants, in perception of their status and possibilities, in their attitude towards technological innovations, or in the activities of worker organisations.

Careful analysis of these findings enables us to explain them not by the impact of technology but by such basic features of the social and economic system as :

(a) social ownership of the means of production;
(b) planned management of the economy;
(c) the high level of social and economic guarantees for the working people;
(d) the leading role of worker organisations in the society;
(e) the highly developed system of participation of the working people in production management.

3.1. Job Security and Technological Change

A most important question connected with automation is the character of its influence on employment. Automated equipment makes it possible to save labour. Does this mean that introduction of such machines should unavoidably result in dismissals and even in unemployment? As a result of socialist transformations, unemployment was completely eliminated in the Soviet Union from the early 1930s. Is there now a threat of emerging unemployment because of automation?

Table 13. Mean score of factors characterising work content given by workers depending on the extent of job satisfaction

	Automated (A) / Non-automated (NA)	Very satisfied	Somewhat satisfied	Partly satisfied, partly not	Somewhat dissatisfied
Variety	NA	5.3	4.7	4.1	2.6
	A	6.5	5.7	5.1	3.8
Independence	NA	6.7	6.1	5.9	5.4
	A	7.0	7.0	7.0	5.3
Responsibility	NA	6.8	6.4	5.8	5.0
	A	7.6	7.4	7.3	6.0
Chance to use your knowledge and training	NA	5.8	5.3	4.3	2.4
	A	7.1	6.2	5.7	3.8
Possibility of working out new and better ways to do your job	NA	4.5	3.7	3.2	3.0
	A	5.4	5.0	4.7	3.0
Need to learn new things	NA	4.6	4.3	3.5	2.6
	A	6.9	5.9	5.3	3.8
Interesting work	NA	6.5	5.4	3.5	2.0
	A	7.9	6.6	4.4	3.8
Chance to develop your own abilities	NA	5.3	4.8	3.6	2.3
	A	7.1	6.0	5.0	4.5

Table 14. The opinion of workers, management, and trade union organisation about the influence of automation on different factors

Factors	Change	Opinion of workers (%)	Opinion of management	Opinion of trade union
Job security	Increase	18	–	–
	No change	77	I,II,III,IV[a]	I,II,III,IV
	Decrease	0	–	–
	Don't know	5	–	–
Worker participation in decisionmaking	Increase	30	II,III,IV	II,IV
	No change	53	I	I,II
	Decrease	2	–	–
	Don't know	15	–	–
Prospects for young workers	Increase	82	I,II,III,IV	I,II,III,IV
	No change	10	–	–
	Decrease	3	–	–
	Don't know	5	–	–
Mental and nervous strain	Increase	63	I,II,III,IV	I,II,III,IV
	No change	20	–	–
	Decrease	13	–	–
	Don't know	4	–	–
Physical strain	Increase	4	–	–
	No change	9	–	–
	Decrease	84	I,II,III,IV	I,II,III,IV
	Don't know	3	–	–
Job satisfaction	Increase	75	I,II,III,IV	I,II,III,IV
	No change	11	–	–
	Decrease	5	–	–
	Don't know	9	–	–
Importance of union membership for workers	Increase	23	–	II,IV
	No change	63	– I,III	
	Decrease	1	–	–
	Don't know	13	–	–

[a] I,II,III,IV plant codes.

Managers of all plants and leaders of all trade union organisations who were interviewed in the course of the investigation unanimously stated that automation does not change job security. The same answer was given by 77 % of the workers interviewed. In fact, 18 % of workers expressed the view that the introduction of automation even increases job security. What is meant in this case is job security at a specific workplace in automated production (Table 14).

The existing level of guaranteed job security was highly valued. In a nine-point rating scale the mean score of automated units was 8.5, while in non-automated units the mean score was 8.6.

Ninety-eight per cent of workers in automated units and 99 % in non-automated units expressed their confidence that the introduction of automation does not threaten them with the loss of their jobs. Ninety-one per cent of those interviewed firmly believed that if their work is taken over by new machines they will be provided with another acceptable job (Table 15).

Table 15. Opinion of workers about providing them with other acceptable work if their present job is taken over by new machines.

(Percentage of the number of respondents)

	If your present job is taken over by new machines do you think you will be provided with other acceptable work?			Total
	Yes	Possibly	No	
Workers of automated production	92.4	6.5	1.1	100
Workers of non-automated production	89.9	8.6	1.5	100
For the whole sample	91.9	7.6	1.3	100

The analysis of answers given by representatives of administration, trade unions, and workers shows that automation itself does not determine the level of employment and job security. The conviction that full employment will be maintained, and that each worker will be provided with work, is connected primarily with confidence in the balanced development of the socialist economy and in the job guarantees which are ensured for all working people.

The employment guarantee is a fundamental factor in the social status of Soviet working people which preconditions their attitudes toward different problems. But it is precisely because of the universal effect of this factor in our country, and because all workers perceive it in the same way, that a correlation analysis is not possible to determine its influence on other social phenomena. In practice, such influence is very great and it is, in particular, being vividly demonstrated by the attitude of workers and their organisations towards technological innovations.

3.2. Attitudes Towards Technological Change

Leaders of all union organisations unanimously and unconditionally stated their approval of technological innovations in their plants. No worker entertained any doubts concerning the subject. Ninety - eight per cent of workers of automated production and 97 % of workers of non-automated units stated their readiness to actively support the introduction of new labour-saving machines and equipment in the plant. The remaining workers said they "quite approved" of these innovations.

When explaining their position in an open-end question, leaders of union organisations noted, first of all, the positive influence of new technology on the improvement of working conditions and on general job satisfaction. On their part, when answering the same open-end question, the largest number of workers singled out such factors as making work easier, increased labour productivity, and more interesting work (Table 16).

It is worth noting that workers mentioned factors which are important not only for them personally but also for the successful operation of the plant. They attach greater importance to labour productivity than to many factors affecting their own personal interests.

It can also be pointed out that, in their answers to these open-end questions, neither union leaders nor workers noted such factors as job security and possibilities for worker participation in decision-making. This is again an example of the impact of general social and economic conditions.

The answers to other closed-questions provide further insight concerning the influence of automation. (Reference should be made once more to Table 14.)

The importance of maintaining full job security has already been noted above. As to worker participation in production management, some managers and leaders of trade union organisations believe that under conditions of automation, worker participation in decision-

Table 16. Reasons given by workers for approving introduction of work-saving new machines and equipment in the plant

(Percentage of respondents in both types of units)

	Making work	Improving productivity	Raising interest of work	Improving working conditions	Improving the quality of product	Raising skill level	Economy of raw materials and energy	Improving work organisation	Higher
Workers of automated units	38.0	24.5	20.7	7.6	3.3	3.8	1.6	0	0
Workers of non-automated units	44.9	19.2	15.2	5.6	6.1	2.0	2.0	2.0	1.5
For the whole sample	41.5	21.9	17.8	6.5	4.7	2.9	1.8	1.0	0.8

Table 17. Ratings given by workers of opportunities provided by their present job

(according to a nine-point scale)

	Automated (A) / Non-automated (NA)	Mean score	Standard deviation	t-Ratio
Job security	A	8.5	1.1	0.2
	NA	8.6	1.2	
Opportunity for up-	A	6.1	2.0	4.6
grading or promotion	NA	5.1	2.1	
Chance to improve	A	6.5	1.8	5.2
occupational skills	NA	5.5	2.2	
Opportunity to getA	7.8	1.5	1.8	
more general education	NA	7.5	1.8	1.8

Table 18. Ratings given by workers to factors affecting upgrading and promotion according to a three-point scale (3 - very important; 2 - somewhat important; 1 - no importance)

	Automated (A) / Non-automated (NA)	Mean score	t - Ratio
Skill	A	2.9	1.901
	NA	2.8	
Working hard	A	2.9	1.178
	NA	2.8	
Seniority in the company	A	2.5	1.111
	NA	2.5	
Being a favourite of the	A	1.1	1.861
boss or supervisor ("pull")	NA	1.2	
Union influence	A	2.4	0.907
	NA	2.3	
Social acitivity	A	2.3	1.006
	NA	2.2	

making undergoes no changes, while others consider that it increases. The same picture emerges from the answers of the workers. Fifty-three per cent said that worker participation remains unchanged, while 30 % saw it as increased. More workers in automated units indicated an increase in this factor.

These two tendencies in opinions are apparently due to different interpretations of the question. If it refers to the possibilities and forms of worker participation in production management, then general social and economic conditions rather than level of automation prevail. If, on the other hand, the question is interpreted as referring to the importance and effectiveness, then workers could mean that the introduction of automation increases both the requirement for active participation of workers as well as the level of general and special education and qualification, which enables workers to be more competent.
Ninety-four per cent of the workers interviewed noted that the introduction of new equipment had, as a whole, affected the workers favourably. This view indicated good opportunities provided by technological change. This should be seen against the background of social and economic factors and, in particular, with possibilities for workers to participate in technological planning.

The information obtained from management and unions, the data from expert evaluations of jobs, and replies of the workers, unanimously point to the fact that the introduction of automated technology is accompanied by higher demands for general education of workers. The possibilities for meeting these demands depend upon social and economic conditions.

An extensive network of higher and secondary evening school and correspondence institutes has been established in the Soviet Union. In all towns and all major plants there are secondary schools which make it possible to raise one's educational level without giving up work. The workers who attend such schools, or higher educational institutions, enjoy numerous benefits under the law.

Opportunities to enhance one's general education are highly appreciated by all workers. In response to a nine-point scale, the mean score of evaluating this factor in automated units was 7.8, while in non-automated units it was 7.5 (Table 17).

The possibilities of improving one's skills and grade and for job promotion depend largely on the type of work. Therefore, workers engaged in automated production units appreciate them more highly (Table 17). At the same time, the implementation of these possibilities is closely linked with the whole system of social production relations and it is not contigent upon technological variations. A case in point is the uniformity of responses of workers both in automated and non-automated units to the question on factors which contribute to improving a person's position (rating) in his department (Table 18). In addition to the importance of skills, hard work, and seniority, workers stress the great role which union support and social activity play. On the contrary, the importance of favouritism ("pull") is flatly denied by most workers.

The data of the study support the fact that relations between people at work are basically determined by the general character of production relations existing in the country. This fact does not mean that the level of technology generally has no effect on relations between people during the work process. The responses of the workers show, for instance, that those engaged in automated production see far more possibilities for assisting one another when doing work and for talking with each other during the work process. The *t*-value of the mean score discrepancy between the evaluation of these factors in automated and non-automated production units is 6.27 and 4.92 respectively. The smallest difference is in personal interest towards one's co-workers (t=2.15). The high level of such interest is generally characteristic of the social and psychological atmosphere at Soviet plants (Table 19).

Responses to other questions also give an idea of the qualitative characteristics of interpersonal relations at the plants, yet these do not bring out any substantial differences between workers in the two types of production units. Informal contacts among workers and between them and management are rather frequent both in the plant and outside (Table 20).

Table 19. Ratings given by workers to relations between co-workers

	Mean Score			
	In total sample	In automated units	In non-automated units	t - Ratio
1. Chance to talk with each other while working	5.08	5.65	4.56	4.92
2. Help each other in getting work done	7.18	7.74	6.67	6.27
3. Take a personal interest	7.19	7.36	7.04	2.15

Table 20. Ratings given by workers to the frequency of informal contacts between the people in the plant according to a three-point scale
(3 - regularly; 2 - occasionally; 1 - never)

	Automated (A) / Non-automated (NA)	Mean score	t - Ratio
Informal talks between workers	A NA	2.6 2.6	0.127
Informal talks of workers with supervisory and administrative staff	A NA	2.3 2.2	0.373
Workers and supervisors have meals or snacks together at the plant	A NA	2.9 2.9	0.195
Joint leisure time activitites of workers, supervisors, and administrative staff	A NA	2.6 2.6	0.616
Invitation of workers by other workers to their home	A NA	2.4 2.4	0.896
Invitation of workers by supervisors and administrative staff to their home	A NA	1.6 1.5	1.274

Opportunities for such informal contacts are enhanced considerably by joint participation in the activities of worker organisations. Workers, technicians, managers, and employees are members of the same union and Party, Komsomol, and other organisations. They associate with each other in the course of general meetings and during the work of various committees and commissions. Shared use of canteens, collective enjoyment of leisure time, and mutual invitations to each others' homes are typical. It is manifest that most workers do not draw a sharp distinction between work and personal life.

Unquestionably, the frequency of informal contacts of workers with one another is higher than that between workers and managerial staff. This is shown especially in mutual visits to homes when comradely relations develop into close friendly ties. At the same time the perception of opportunities to criticise a foreman is also related to the perception of opportunities to criticise the top echelon. Of course, social and professional distinctions, differences in educational level, and specific interests also remain important. But, on the whole, the data obtained convincingly demonstrate absence of any aloofness on the part of managerial personnel.

Numerous studies by Soviet sociologists have been devoted to studying interpersonal relations at industrial plants.[12] The sociological data reflect a high level of informal contacts between workers, technicians, engineers, and managers of the plants. This is supported by the data on social origin of plant managers. At the Pervouralsk Novotrubny Plant, for instance, 42 % of technicians and engineers come from workers, 32 % from peasants, and 26 % from clerical workers.[13]

Nearly 90 % of the workers interviewed said that managers and administrators of a plant do not look down on them because they are workers. These responses characterise the overall atmosphere at Soviet plants where a worker's status is very high and where any disdainful attitude towards workers would be harshly decried by the members of the entire work team and by the workers' organisations.

When an individual worker does feel some condescension on the part of managers, this may indicate either the wrong method of management or the wrong reaction of some workers to well-founded comments and requirements made by managers. Public sanction and open criticism within a group serve as an effective means of dealing with such behaviour.

Broad opportunities for open criticism are an integral element of Soviet democracy. The system of worker participation in production management could not be effective if the workers' right to criticise was limited or if they feared harassment because of it.

The data obtained from the study show that the overwhelming majority of workers regard it as possible to openly criticise their workmates, including foremen, shop engineers, shop chiefs, and plant managers (Table 21). There is no significant difference in responses to this question given by workers in automated and non-automated units, since the opportunities for and level of criticism are determined by social conditions at a plant and not by the type of technology.

The small percentage of workers who do not see opportunities for open criticism of their managers reflects subjective views of the respondents rather than objective conditions. There are workers who do not want to mar their relations with managers or who do not consider themselves sufficiently competent to make critical statements.

Not all workers, however, who criticise rank-and-file workers consider it possible to make a similar criticism of engineers, heads of departments, and plant managers. At the same time the perception of opportunities to criticise a foreman is also related to the perception of opportunities to criticise a foreman is also related to the perception of opportunities to criticise the top echelon. This clearly reflects the growing social activity of workers where ascent to each step opens up new horizons and makes it possible to keenly respond to wider problems of running a unit, department, or plant. The first step towards such development is the social and psychological willingness of workers to make publicly a critical judgement of their foreman's work.

Table 21. Opinion of workers about the possibilities of criticising in public and in the presence of persons concerned
(Percentage of respondents in both types of units)

Categories of persons concerned	Is it possible to criticise them in public?			
	Automated units		Non-automated units	
	Yes	No	Yes	No
Rank-and-file workers	100	0	99	1
Direct supervisor (foreman)	97	3	99	1
Shop engineers	93	7	91	9
Head of the department	92	8	90	10
Plant managers	86	14	83	17

Opportunities for public and objective criticism characterise in many respects the social atmosphere at a plant. The correlation analysis revealed, inter alia, a significant negative correlation between this factor and influence seeking to get a promotion as well as managers' condescending attitude towards workers (see Chart 1).

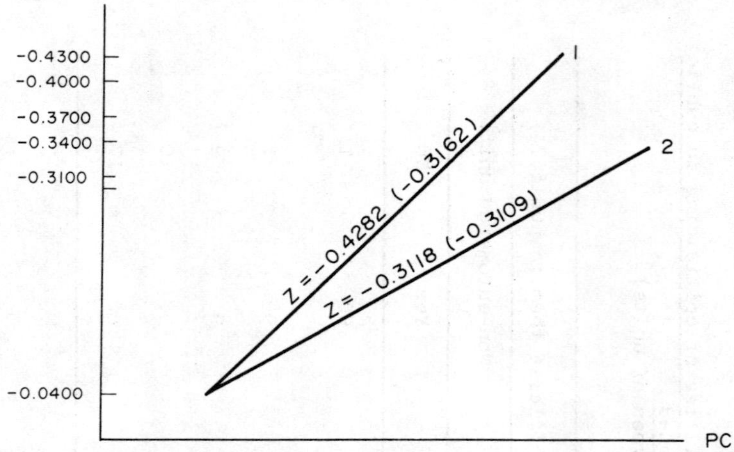

Chart 1. Correlation between possibilities for openly criticising management (PC), favouritism ("pull") (1), and condescensionon the part of management (2). Data for the head of the department without brackets; for plant managers in brackets.

The great importance attached to this factor by workers, foremen, shop superintendents, chief engineers, directors, Party, union, and Komsomol leaders was supported by the sociological study which had been carried out at fifteen plants in Leningrad, Moscow, Kharkov, and Aktubinsk, and which covered 5000 respondents.[14]

The highest Party and union bodies ensure the development of businesslike criticism and self-criticism. Thus, for example, a check-up conducted by the CPSU Central Committee in the Tambov region revealed serious shortcomings in the state of criticism at a number of plants. Some managerial workers, while agreeing with critical remarks, did nothing to rectify the shortcomings. There were ascertained cases of suppression of criticism, harassment for critical statements, and cases where those responsbile were not reprimanded.

The CPSU Central Committee adopted a resolution on "The State of Criticism and self-criticism in the Tambov regional Party organisation",[15] calling special attention of the respective leaders to the need of adherence to the principle of creating conditions for candid criticism from below and taking effective measures in response to criticisms. The CPSU suggested dealing decisively with cases of suppression of criticism and, if necessary, punishing even to the extent of removing from their posts those managers who react in the wrong way to criticism, take just remarks addressed to them as downgrading their reputation and authority, put personal egotism above social interests, and avenge themselves on those who make such criticism.

The discussion of that resolution in worker organisations of the entire country contributed to the growth of activities by worker masses and to the strengthening of democratic principles of management.

3.3. Union activities

Does automation per se influence the activities of trade unions? Leaders of two plant unions responded that the importance of the union for workers does not change due to automation. In the view of leaders of two other union organisations, this importance increases. The workers themselves gave different answers to that question. Most of them (63 %) believe that the importance of trade unions does not change, while 23 % think that it is increasing (Table 14).

As a matter of fact, the responses to that question, as well as to the question of the influence of automation on worker participation in management, reflected various aspects of the problem. As to the role of unions and their functions, these are determined not by technology but by the general status of these organisations in society. But if it is a question of concrete tasks, new problems arise due to automation which causes additional attention on the part of unions to solve them.

According to the study, the overwhelming majority of workers regard as very important every direction of union activities linking closely their personal interests with public ones, in particular with the tasks of improving labour productivity. This is shown by the correlation analysis (Table 22).

Job security ranked last in the workers' evaluations of the importance of union activities. This again reflects the belief that job security is assured by the social and economic system. It is significant that the correlation analysis shows that the tasks of unions in assuring job security are most closely linked with the influence of unions on the overall social and economic policies of the country.

Worker satisfaction with certain aspects of union activities depends, of course, on the effectiveness of the work of specific union organisations. But this effectiveness is in many ways predetermined by the status of the union, its rights, and its possibilities.

In analysing the responses of the workers about the attention the union pays to the various different activities, no substantial difference was found in the workers' evaluation between the two selected types of production units.

The tremendous influence of general social and economic factors upon the processes and social consequences of automation does not imply that these factors are operating automatically, thus ruling out the need for conscientiously regulating human activity. On the contrary, in order to utilise as fully as possible objectively favourable possibilities in the interests of the entire society and every worker, a great amount of creative work is called for by governmental, non-governmental bodies, economic managers, worker organisations, and scientists. This was the theme referred to by the Secretary General of the CPSU Central Committee, L. I. Brezhnev, at the 24th Congress of CPSU, who stressed the currently central task of historic importance". . . . to organically fuse the achievements of the scientific and technical revolution with the advantages of the socialist economic system, to unfold more broadly our own, intrinsically socialist forms of fusing science with production."[16]

Table 22. Ratings Given by Workers to the Following Issues in Union Activities[a]

Items	1	2	3	4	5	6	7	8	9
1. Improving wages, pensions, etc., of workers	I								
2. Developing cultural and educational activities in the plant	0.28	I							
3. Providing recreational facilities (including vacations)	0.25	0.41	I						
4. Improving productivity	0.34	0.31	0.40	I					
5. Awakening a strong political attitude and consciousness among the workers	0.37	0.40	0.36	0.51	I				
6. Improving the terms and conditions of work	0.34	0.53	0.25	0.32	0.39	I			
7. Influencing the general social and economic policy of the country	0.41	0.32	0.32	0.30	0.52	0.37	I		
8. Handling worker grievances (complaints) and disputes	0.27	0.25	0.19	0.24	0.26	0.31	0.36	I	
9. Ensuring job security for workers	0.30	0.15	0.16	0.12	0.26	0.20	0.37	0.15	I
10. Changing work organisation to increase job satisfaction	0.29	0.12	0.20	0.34	0.16	0.30	0.15	0.20	0.25

[a] The table of correlation coefficients r on the items. A correlation of 0.17 is significant at the 0.01 level (meaning that it would occur by chance only once in a hundred).

NOTES

1 The question about taking decisions to cut back production or close down a plant was the only exception. Neither workers, management, nor unions of the investigated plants have ever encountered such a situation in real circumstances. Therefore, 65.2 % of the workers interviewed replied "don't know" to that question, while the rest answered on the basis of their hypothetical ideas alone. Union and management representatives could not answer the question at all.

2 V. I. Mouchatchev and V. S. Borovik, Working Class in Industrial Management, (in Russian), Moscow, pp. 31 and 47.

3 V. N. Uspensky, Improvement of forms of worker participation in production management in conditions of advanced socialism (in Russian), Scientific Reports of Advanced Schools. Scientific Communism 1975 (1), pp. 35-36.

4 A. O. Razhigaev, Creative productive activity of workers" (in Russian), Kommunist 1973 (17), p. 69.

5 V. N. Uspensky, op. cit., p. 35.

6 These indices (marked with an asterisk in the text) are explained in the Appendix.

7 L. S. Bliachman and O. I. Shkaratan, Scientific and Technological Revolution: The Working Class and Intelligentsia (in Russian), Moscow, 1973, pp. 194-195.

8 The Working Class and Technological Progress (in Russian) (M. T. Yovchuk and L. N. Kogan, (eds.) Moscow, 1965; Mental World of the Soviet Worker: An Experience of Concrete Sociological Research (in Russian), Moscow, 1972.

9 Social Problems of Work and Production (in Russian), Moscow-Warsaw, 1969, p. 126.

10 A. G. Zdravomyslov et al. (eds.), Man and His Work, White Plains, IASP, 1970, p. 289.

11 Relations of production are relations among men in the social production of their existence. The main element of these relations is property relations, i.e. people's relation to the means of production. "The totality of these relations of production constitutes the economic structure of society, the real foundation, on which arises a legal and political superstructure and to which corresponds definite forms of social consciousness" (K. Marx, A Contribution to the Critique of Political Economy, London, 1971, p. 20.)

12 S. P. Syrotkin, Evolution of Production Collective Team, Moscow, 1973; G. L. Smirnov, Soviet Man, Moscow, 1972; V. I. Zatsepin, Interpersonal Relations in a Collective Team, 1970; Ivanov and Frish, Your Production Collective Team, Moscow, 1970; G. A.

Antipina, Study of small Groups in Sociology and Social Psychology, 1967. Moulding of personality in a production collective team, Moscow, 1968; O. I. Zotova, Experience of studying mutual relations in worker collectives (Reports of the Second Congress of Psychologists Society), Moscow, 1963 (all in Russian).

[13] Mental World of the Soviet Worker, (loc. cit.), p. 18.

[14] I. I. Changly, Labour, (in Russian) Moscow, 1973, p. 455-456.

[15] Kommunist, 1975 (No. 4).

[16] Documents of the 24th Congress of the CPSU, Novosti Press Publishers, Moscow, 1971, p. 69.

Appendix

Supplement to Table 6

Formulae for calculation of composite indices for work content and working conditions on the basis of Research Instrument IV - Work Place Observation (see Part III of this book).

Notes: Z_m is the numerical value of index m.
X_n is the mean score for n-item of RI-IV.
(n is a number of the item in RI-IV).
T, V is the type of scale used for the calculation of index
(T is the time scale; V is the scale of variance).

I. Job Variety

$$Z_1 = \frac{X_{1T}.X_{1V}+X_{2T}.X_{2V}+X_{3T}.X_{3V}+X_{4T}.X_{4V}+X_{5T}.X_{5V}}{5} \cdot \frac{1}{X_{50}}$$

II. Responsibility for Working Process

$$Z_2 = X_{52}$$

III. Level of Automation

$$Z_3 = \frac{X_{72}+X_{73}+X_{74}+X_{83}+X_{84}+X_{75}+X_{76}+X_{77}+X_{80}+X_{81}+X_{85}}{11} \cdot \frac{X_{78}+X_{79}+X_{82}}{3}$$

IV. Job - Required Skills

$$Z_4 = \frac{X_{10}+X_{11}+X_{12}+X_{56}}{4}$$

V. Information of Processing Activities

$$Z_5 = \frac{X_7+2.X_8+3.X_9}{3}$$

VI. Job - Required Communications

$$Z_6 = \frac{\frac{1}{X_{67}}+X_{68}+X_{69}+X_{70}+\frac{1}{X_{71}}}{5}$$

VII. Use of Devices and Equipment

$$Z_7 = \frac{X_{13}+X_{14}+X_{15}+X_{16}+X_{17}+X_{18}+X_{19}+X_{20}+X_{21}}{9}$$

VIII. Sources of Job Information

$$Z_8 = \frac{X_{1T}+X_{2T}+X_{3T}+X_{4T}+X_{5T}}{5}$$

IX. Discomfort of Physical Working Conditions

$$Z_9 = \frac{X_{55}+X_{57}+X_{58}+X_{59}+X_{60}+X_{61}+X_{62}+\frac{1}{X_{63}}+X_{64}+X_{65}+X_{66}}{11}$$

X. Structure of Job

$$Z_{10} = \frac{\frac{1}{X_{67}}+X_{46}+X_{49}+X_{53}+X_{54}}{5}$$

XI. Physical Strain

$$Z_{11} = \frac{5.X_{27}+X_{28}+X_{29}+X_{30}+X_{31}+X_{32}+X_{33}+X_{34}+X_{35}+X_{36}}{10}$$

$$\frac{X_{22}+X_{23}+X_{24}+X_{25}+X_{26}+X_{37}+X_{38}+X_{39}+X_{40}+X_{41}+X_{42}+X_{43}+X_{44}+X_{45}}{14}$$

XII. Monotony of Job

$$Z_{12} = \frac{X_{47}}{X_{48}}$$

XIII. Intensity of Job

$$Z_{13} = \frac{X_{51} + X_{52}}{2} \cdot \frac{Z_{11}}{X_{48}}$$

PART III

1. Research Instruments

Research Instrument I

SOCIAL AND ECONOMIC SYSTEM OF THE COUNTRY AND ITS INFLUENCE ON TECHNOLOGICAL CHANGES AND THEIR SOCIAL CONSEQUENCES

GUIDELINES FOR THE COLLECTION OF GENERAL INFORMATION
(Country and Branch of Industry under Investigation)

The purpose of this general information is to provide preparatory and supplementary data for the central concern of the study, i.e. of comparing among various countries the effect of modern technology upon industrial workers.

Its importance results from the hypothesis that a country's political, economic, and social systems will exert a significant influence upon the impact of automation.

The investigator should collect relevant data from publications and documents such as government statements, union programmes, etc., and from people with specialised knowledge.

Questions requiring quantitative data should be answered on the basis of ILO and United Nations statistical year books, supplemented where necessary and possible by national official statistics. Source of statistical data must be shown in every case. The branch of industry under investigation must be specified in every case (e.g. automobile industry, metal industry, truck industry, etc.).

Other questions should be answered in short verbal form, leaving it open to each national group to provide more detailes characteristics of the social and economic system of its country.

The written report on general information should be as concise and precise as possible, with considerable freedom to individual national study groups to utilise country-specific materials.

This information is to be viewed primarily as non-comparable, descriptive material that would provide a basic understanding of the context within which each national study was conducted.

GENERAL STRUCTURE OF THE RESEARCH INSTRUMENT

Part I : Basic features of social and economic system of the country

Part II: The structure of power and process of industrial development

Part III : System of employment and income security for industrial workers

Part I: BASIC FEATURES OF SOCIAL AND ECONOMIC SYSTEM OF THE COUNTRY

1. Types of ownership of the means of production.
   (a) Give a brief description.
   (b) Give the percentage of gross national product accounted for by each of the following sectors in 1973:

      (i) private sector;
      (ii) state or social sector;
      (iii) other sectors.

   (c) Give the percentage of total output in this branch of industry accounted for by each of the following sectors in 1973:

SHARE IN:	Private	State or social	Other
Total output			

2. Development of the economy.
   (a) Brief description of the development during the past 20 years.
   (b) Rate of growth of national economy (as a percentage of 1955).

Years	1955	1960	1965	1970	1971	1972	1973
Gross national product	100%						
National income	100%						
Total industrial output	100%						

   (c) Rate of growth in the branch of industry under investigation in units produced (as a percentage of 1955).

Years	1955	1960	1965	1970	1971	1972	1973
Passenger cars	100%						
Trucks	100%						
Tractors	100%						

(d) Briefly describe the development of automation over the last 20 years, using quantitative data wherever possible. Data should refer to the country as a whole, branch of industry under investigation.

3. Social stratification (or class structure). Describe briefly.

4. Description of type of economy.
Planned.
Market.
Mixed.
Other.
Please explain what the term you have chosen menas in your country.

5. Structure of employment.
(a) Percentage of the total economically active population employed in the following sectors for: 1955, 1960, 1965, 1970, 1971, 1972, 1973.

Agriculture	Commerce
Mining,quarrying, manufacturing	Transport, storage, communication
Construction	Other
Electricity, gas, water	Total economically active population.

(b) Percentage of the total economically active population which is employed in the branch of industry under investigation for: 1955, 1960, 1965, 1970, 1971, 1972, 1973.

6. Rate of unemployment in the country for: 1955, 1960, 1965, 1970, 1971, 1972, 1973.

7. System of education and job training. Give a brief description of:

(a) National educational policy and structure of education system.
(b) System of general education.*
(c) System of technical education.*
(d) System of occupational education, job training and retraining with special reference to technological change:
   (i) brief description;
   (ii)national and branch of industry data (where available).
(e) Types of additional work-oriented education (adult development courses, trade union courses, etc.)

* For 7b and 7c give.

(i) attained levels of education as percentage of population in 1950 and 1970;

(ii) enrolement at various levels as percentage of eligible (corresponding) age groups during the most recent years.

PART II: THE STRUCTURE OF POWER AND PROCESS OF INDUSTRIAL DEVELOPMENT

1. The role of national (state) governing bodies (legislative, executive, judicial) in:

(a) Regulating the national economy.
(b) Influencing technological change.
(c) Influencing the social consequences of technological changes.

2. The role of private companies and associations in:

(a) Influencing the national economy.
(b) Influencing technological change.
(c) Influencing the social consequences of technological changes.

3. The role of political parties in:

(a) Influencing the national economy.
(b) Influencing technological change.
(c) Influencing the social consequences of technological changes.

4. The role of trade unions (for the country and branch of industry).

(a) Give a brief description of the structure of trade union organisation in the country and the branch of industry under investigation.
(b) Indicate the extent of the union organisation in the country (measuring trade union membership either as a percentage of total wage and salary workers or if possible as separate percentages for blue and white-collar workers for: 1955, 1960, 1965, 1970, 1971, 1972, 1973.
(c) Indicate the percentage of union membership in the branch of industry either as a percentage of total wage and salary workers or if possible as separate percentages for blue and white-collar workers for: 1955, 1960, 1965, 1970, 1971, 1972, 1973.
(d) Describe briefly the ways by which unions influence social and economic policy:

   (i) by participation in national economic and social planning councils, boards, etc.;
   (ii) through political action (supporting political candidates, participating in national political parties, etc.);
   (iii) other (describe).

5. The general systems of worker participation:

(a) In decisionmaking (on national, branch of industry, and plant levels) - and its subject - matters and forms.
(b) In suggestion systems.
(c) In voluntary societies concerned with technological change.

6. Special country-specific features which characterise the workers' attitudes toward work (describe very briefly).

7. Regulation of wages and working conditions. Give brief description of:

(a) Legislation.
(b) Collective agreements.
(c) Regulations at the plant level.

8. Methods of solving collective industrial disputes:

(a) General description of methods.
(b) Number and duration of strikes:

(i) In the country for: 1955, 1960, 1965, 1970, 1971, 1972, 1973
Number of strikes. Number of workers involved. Duration (man-days lost/man-days worked)
(ii) If possible, in the branch of industry being investigated for: 1955, 1960, 1965, 1970, 1971, 1972, 1973.
Number of strikes. Number of workers involved. Duration (man-days lost/man-days worked).

PART III: SYSTEM OF EMPLOYMENT AND INCOME SECURITY FOR INDUSTRIAL WORKERS

1. Job security
(a) Guarantees for employment
(b) Protection against dismissals and redundancy provided by legislation and collective agreements.

2. Wage guarantees (during employment):
(a) Systems of wage classification (grades).
(b) Guarantees of attained wage level.
(c) Payments for time not worked (e.g. trade union activity, machine stoppages, etc.)

3. Working hours
(a) Normal (standard) working hours per week.
(b) Regulation of overtime work.
(c) Regulation of work under less-than-normal (standard) working hours.

4. System of governmental (state) and trade union control over safety, health, and working conditions.

5. The main characteristics of the country's social security system. Give brief description, including:
(a) Age of retirement on an old age pension.
(b) Amount of old-age pension as compared with a worker's wage (percentage analysis).
(c) Period of waiting and terms of payment in case of:
(i) injury;
(ii) sickness.
(d) Amount of
(i) injury, and
(ii) sickness pay as compared with a worker's wage (percentage analysis.
(e) Period of waiting and terms of payment of unemployment compensation.
(f) Amount of unemployment compensation as compared with a worker's wage (percentage analysis).

(g) System of social insurance financing (all types of insurance).

Contribution of workers	Contribution of employers	Contribution of state
(percentage of wage)	(percentage of a worker's wage. If not applicable state other basis for comparison)	(percentage of expenditure of total social insurance services)

(h) Amount of additional expenditures of workers on medical services in case of temporary disability.

6. The existing procedures for solving individual grievances.

Research Instrument II

INFORMATION OBTAINED FROM MANAGEMENT AT THE PLANT LEVEL

INSTRUCTIONS FOR INTERVIEWERS

1. This research instrument is divided into three parts:

   Part I: Background information.
   Part II: Questionnaire for managers.
   Part III: Supplementary information.

2. Part I: Background Information

   This part of the Research Instrument is to find out largely statistical data about the plant so as to provide background context for the research. These statistical data should be obtained from available records and documentation. These data may not be immediately available and require some inquiry. Therefore it is envisaged that this part of the Research Instrument may be left with top management of the plant for completion following the interview required by Part II: Questionnaire for managers.

3. Part II: Questionnaire for Managers

   This part of the Research Instrument must be administered through personal interviews. The interviews are to be conducted with top managers of the plant.

4. PART III: Supplementary Information

   It is expedient to enable the respondent(s) to supply extra information and any data not provided for directly in the Research Instrument, if these data, in the opinion of the respondent, are of importance for the problems under investigation. Extra information may be given at the time of completing Parts I and II or subsequently. These data should be recorded in supplementary information.

5. For the purpose of this study we are investigating:

(a) the transfer line unit or other specific technological operation and all jobs dependent on the automated equipment, as compared with

(b) similar machining operations on manually controlled machine tools within the same plant.

PART I : BACKGROUND INFORMATION

1. What type of ownership applies to this company? (Check one.) State or social ownership. Private ownership. Other (please explain).

2. How long has this plant been in existence?
Less than 10 years. 10-20 years. More than 20 years.

3. What are the main products of this plant?

4. Indicate the approximate present total employment in the plant.
Less than 1000. 1000 - 5000. More than 5000.

5. Indicate the exact number of employees in the two units under investigation:

(a) Automated.
(b) Nonautomated.

6. How long ago was the transfer line under investigation installed in the plant?
Less than 1 year. 1-3 years. 3-6 years. 6-10 years. More than 10 years.

7. (a) If there is a suggestion system, indicate the number of suggestions made in 1973 and the degree to which they were accepted and implemented:
   (i) in the automated unit;
   (ii)in the nonautomated unit.
   (b) If there were substantial differences between the two units, please explain the reasons.
   (c) Who decides to accept or reject suggestions?

8. Which are the present wage payment systems in the automated and nonautomated units under investigation?

   Time-rate. Piece-rate. Other (please specify).

9. What are the normal (standard) hours per week in the plant?

10. (a) What was the average overtime per man per week in each unit under investigation during 1973?
   (i) Automated;
   (ii)nonautomated.
   (b) If there were any substantial differences in this respect between the automated and nonautomated units, please explain.

11. (a) Where there any reductions from standard working hours in each shift under investigation during 1973? (Yes, No).
   (b) If there were any substantial differences in this respect between the automated and nonautomated units, please explain.

12. How many shifts do you normally operate at the plant in the automated unit and in the nonautomated unit?

13.(a) Are there any differences in the frequency and duration of regular official breaks during the working day between the automated and nonautomated units under investigation? (Yes, No).
(b) If yes, please explain.

14. With respect to automation, does your plant have any of the following provisions? (Yes, No).

Protect against dismissal
Provision of alternative job opportunities
Provision of training facilities
Compensation for dismissed workers
Compensation for transferred workers
Other (please specifiy).

15.(a) During 1973 were there any substantial differences between the two units under investigation with regard to:
Automated unit
Higher Lower No difference
(i) Rate of labour turnover
(ii) Rate of unauthorised absences
(iii) Accident rate (according to the classification system you use)
(iv) Rate of formal (official) individual grievances
(b) If there were substantial differences in the rate of labour turnover, please explain the reasons.
(c) If there were substantial differences in the rate of unauthorised absences, please explain the reason.
(d) If there were substantial differences in the accident rate, please explain the reasons.
(e) If there were substantial differences in the rate of formal (official) individual grievances, please explain the reasons.

P A R T II: QUESTIONNAIRE FOR MANAGERS

Interviewer complete:

(1) Name of respondent (not obligatory)
(2) Position held in the plant.

1. How important for your plant are questions connected with automation?
(Not important, somewhat important, very important)
(a) At the present time.
(b) In the next 5 years.

2. To what extent has your plant been concerned with the following problems of automation? (Not at all, Somewhat, Much, Very much.)

Transfers
Dismissals (redundancies)
Retraining and improvement of skills
Working hours and shift work
Wage payment system
Wage level
Working conditions
Job satisfaction
Relations between people at work.

3. What were the reasons for installing more advanced machines and equipment in the plant? (Check those reasons which apply.)
To what extent were these aims achieved? (Fully, Partly, Not at all.)

To replace wornout machines and equipment.
To overcome shortage of workers.
To reduce labour costs per unit of output.
To reduce capital costs per unit of output.
To reduce the work load of the personnel.
To make work places safer.
To improve the working conditions.
To improve the quality of the product.
To expand production.
To meet competition.
To increase the prestige of the plant.
Other reasons: specify.

4. (a) Is it the usual practice in the plant for workers to be informed prior to the introduction of new, more advanced machines and equipment? (Yes, No.)
(b) If yes, by what means? (Yes, No.)
Through worker participating in long-term planning for technological innovations.
Through official announcement by management.
Through informal communication from management.
Through the union, worker councils, self-management committees, etc.
(c) If yes, how long before the installation is the information given?
Less than 1 month, 1-3 months, 3-6 months. More than 6 months.

5. How are decisions on the following matters made at the plant?
Management takes decisions:
unilaterally
after preliminary discussions with workers' representatives
only with consent of workers' representatives

1. Development of production plan.
2. Introduction of new machinery and equipment.
3. Work organisation and working conditions.
4. Personnel planning (planning of future work force demands).
5. Determination of basis and method of wage payment.
6. Worker job classification and pay scale.
7. Determination of bonuses and other incentives.
8. Hiring of workers.
9. Promotion of workers.
10. Selection of workers for training courses.
11. Transfer to other work places.
12. Allocation of overtime work.
13. Disciplinary measures.
14. Lay-off and discharge of workers.
15. Cut-back of production and closing down of plant.
16. Recruitment and dismissal of managerial staff.

6. What organisations participate in making decisions on each of the above-mentioned matters? (Please indicate by numbers from question 5.)

   Trade union

   Worker council

   Joint labour-management committees (works council)

   Other (please specify)

7. Are there plans for technological changes in the plant during the next 5 years? (Yes, No.)

8. What effect has participation by worker representatives regarding the above-mentioned matters on the task of management? (Makes it easier, Makes no difference, Makes it more difficult.)

9. By what means is it usual for workers to put forward their own ideas on technical improvements?

   Proposals at labour-management technical committees.
   Suggestion system.
   Proposals at workers' meetings.
   Self-management committees.
   "Club of innovators".
   Informal discussion with foreman.
   Others (specify).

10. In your opinion, how does automation generally influence the following items (Increase, No change, Decrease, Don't know)

    Level of employment.
    Opportunities for promotion/ upgrading.
    Wages.
    Social status of the worker.
    Level of occupational skills.
    Hours of work (including overtime).
    Occupational safety and health.
    Physical strain.
    Date of interview.
    Name of interviewer.
    Mental and nervous strain.
    General educational requirements.
    Job training requirements.
    Job satisfaction.
    Participation of workers in decisionmaking.
    Job security.
    Union membership.
    Prospects of young workers.

P A R T III: SUPPLEMENTARY INFORMATION

Date of interview.
Name of interviewer.

Research Instrument III

INFORMATION OBTAINED FROM TRADE UNIONS AND OTHER WORKER REPRESENTATIVES

INSTRUCTIONS FOR INTERVIEWERS:

1. This research instrument is divided into four parts:
   - Part I: Questionnaire for trade union leaders in the branch of industry under investigation.
   - Part II: Supplementary information from trade union leaders in the branch of industry under investigation.
   - Part III: Questionnaire for trade union leaders and other worker representatives at the plant level.
   - Part IV: Supplementary information from trade union leaders and other worker representatives at the plant level.

2. Parts I and III: Questionnaires

   There are two questionnaires contained in this research instrument. These must be administered through personal interviews. They must be used as follows:

   Part I is the questionnaire to be used at the national level. It is to be administered to trade union leaders in the branch of industry under investigation. It may be administered to one or more persons, depending on the national situation.

   Part III is the questionnaire for use at the plant level.

3. Parts II and IV: Supplementary Information

   It is expedient to enable the respondents to supply extra information and any other data not provided for directly in the research instruments if these data, in the opinion of the respondents, are of importance for the problems under investigation. Parts II and IV are for the recording of this information; Part II for national level; Part IV for plant level.

4. Where statistical data are required, these should be obtained from available records and documents.

5. By "union" we mean those unions which officially represent the workers in the industry under investigation.

6. Interviews should include leaders of the various unions and/or other worker representatives related to the units under investigation. Separate questionnaires should be administered to each of these individuals.

7. For the purpose of this study we are investigating:

(a) the transfer line unit or other specific technological operation and all jobs dependent on the automated equipment, as compared with
(b) similar machining operations on manually controlled machine tools within the same plant.

QUESTIONNAIRE FOR TRADE UNION LEADERS IN THE BRANCH OF INDUSTRY UNDER INVESTIGATION (NATIONAL LEVEL)

(This questionnaire may be administered to one or more persons, depending on the national situation.)

Interviewer complete:
Name of respondent. Name of union. Position(s) held in union.

1. What are your major duties within this union?
2. At the present time what are the most important issues with which your union is dealing?
3. How important to your union are questions connected with automation? (Not important, Somewhat important, Very important, Most important.)
(a) At the present time.
(b) In the next 5 years.

4. In your opinion how does automation generally influence the following items? (Increase, No change, Decrease, Don't know)

Level of employment
Opportunities for promotion/upgrading
Wages
Social status of the worker
Level of occupational skills
Hours of work (including overtime
Occupational safety and health
Physical strain
Mental and nervous strain
General educational requirements
Technical educational requirements
Job training requirements
Job satisfaction
Participation of workers in decisionmaking
Job security
Union membership
Prospects for young workers

5. To what extent has your union been concerned with the following problems of automation? (Not at all, Somewhat, Much, Very much).

Transfers
Dismissals (redundancies)
Retraining and improvements of skills
Working hours and shift work
Wage payment system
Wage level
Working conditions
Job satisfaction
Relations between people at work
Other (specify)

6. (a) What is your union's attitude toward labour-saving machines and equipment in your industry? (Check one.)

   Active support — Reservation
   Co-operation — Opposition

   (b) Please explain your reasons.

7. What actions have been taken by your trade union concerning technological change? (Yes, No, No answer.)

   Resolutions, statements, and declarations
   Publications
   Informal discussion with employers
   Negotiations with employers
   Appeal to the government or governmental bodies
   Internal activities, e.g. organisation of study groups, setting up courses on automation, establishing training facilities, etc.
   Other (specify)

8. Have your union members been involved in industrial disputes arising from problems of technological change in the period 1968-1973?

   None
   Official strikes (give number)
   Unofficial strikes (give number).
   Other acts of dispute (describe what these acts were)

   Date of interview
   Name of interviewer.

SUPPLEMENTARY INFORMATION FROM TRADE UNION LEADERS IN THE BRANCH OF INDUSTRY UNDER INVESTIGATION

Date of interview.
Name of interviewer.

QUESTIONNAIRE FOR TRADE UNION LEADERS AND OTHER WORKER REPRESENTATIVES AT THE PLANT LEVEL

Note for interviewer:
whenever the term "union" is used, change if necessary to the name of the appropriate worker organisation.

Interviewer complete:

Name of respondent.
Name of the union (other worker organisation).
Position(s) held by the respondent.

1. What are your major duties within the union?

2. How many unions exist in your plant?

3. (a) What percentage of the total work force in the plant are union members?
   (b) What percentage of the total work force in the plant are members of your union?

4. What are the issues of highest current importance for your union in its dealing with management in your plant?

5. How important to your union are questions connected with automation?

   Not important. Somewhat important. Very important.

6. In your opinion, how does automation generally influence the following items? (Increase, No change, Decrease, Don't know.)

   Job security
   Participation of workers in decisionmaking
   Prospects for young workers
   Physical strain
   Mental and nervous strain
   Job satisfaction
   Importance of unions for workers

7. To what extent has your union been concerned with the following problems of automation? (Not at all, Somewhat, Much, Very much.)

   Transfers
   Dismissals (redundancies)
   Retraining and improvements of skills
   Wage payment system
   Wage level
   Working conditions
   Job satisfaction
   Relations between people at work
   Other (specify)

8. Which worker organisations in the plant deal with problems related to automation?

9. (a) If labour-saving new machines and equipment are introduced into your company would your union

   (i) approve of such changes,
   (ii) disapprove of such changes? (Check one.)

   (b) Please explain why.

10. Is your union aware of plans for technological changes in the plant during the next 5 years? (Yes, No.)

11. With respect to automation within your plant how satisfactory are the following items concerning protection of workers? (Unsatisfactory, Fairly satisfactory, Very satisfactory.)

   Protection against dismissal
   Provision of alternative job opportunities
   Provision of training facilities
   Compensation for dismissed workers
   Compensation for transferred workers
   Other (specify)

12. How are decisions on the following matters made at the plant?

   <u>Management takes decisions</u>:

Unilaterally	After preliminary discussion with the workers' representatives	Only with consent of the workers' representatives

   1. Development of production plan.
   2. Introduction of new machinery and equipment.
   3. Work organisation and working conditions.

4. Personnel planning (planning of future work force demands).
5. Determination of basis and methods of wage payment.
6. Worker job classification and pay scale.
7. Hiring of workers.
8. Promotion of workers.
9. Selection of workers for training courses.
10. Transfer to other workplaces
11. Allocation of overtime work.
12. Disciplinary measures.
13. Lay-off and discharge of workers
14. Cut-back of production and closing down of plant.
15. Recruitment and dismissal of managerial staff.

13. What organisations participate in making decisions on each of the above-mentioned matters? (Please indicate by numbers from question 12).

Trade union	Joint labour-management committee (works council)
Workers council	Other (specify)

14. (a) Do you think the extent of and methods for participation of the workers' representatives in these questions are sufficient for them to perform their duties effectively?

(i) Sufficient;
(ii) not fully sufficient;
(iii) not sufficient at all.

(b) If above answer is (ii) or (iii), please explain your answer.

15. (a) Is it the usual practice in the plant to inform the union prior to the introduction of new, more advanced, machines and equipment? (Yes, No.)

(b) If yes, by what means?

Through the union participating in longterm planning	Through informal announcement from management.
Through official announcement by management.	Other (please specify).

(c) If yes, how long before the installation is the information given?
Less than 1 month, 1-3 months, 3-6 months. More than 6 months.

(d) Is the length of time sufficient for the union? (yes, No).

(e) Is the amount of information sufficient for the union? (Yes, No).

16. (a) Has the introduction of technological changes during recent years led to any individual grievances between workers and management? (Yes, No).

(b) If yes, were these grievances generally supported by the union? (Yes,No).

(c) If there were grievances due to the introduction of technological change, what were the issues involved?

Wages
Work load
Safety hazards
Transfers
Transfers
Dismissals (redundancies)
Working hours and shift work
Other (specify).

17. (a) Has the introduction of technological changes during recent years led to any collective disputes between workers and management? (Yes, No).

(b) If yes, were these collective disputes supported by your union? (Yes, No).

(c) If there were collective disputes due to the introduction of technological change, what were the issues involved?

Wages
Work load
Safety hazards
Transfers
Dismissals (redundancies)
Working hours and shift work
Other (specify)

Research Instrument IV

WORK PROCESS AND WORK PLACE OBSERVATION

P A R T I : INSTRUCTIONS FOR INVESTIGATOR

For the purpose of this study we are investigating an automated unit- that is the transfer line unit or other specific technological operations and all jobs dependent on the automated equipment.

And a non-automated unit - that has similar machining operations on manually controlled machine tools within the same plant.

INFORMATION TO BE OBTAINED

The Work Process and Work Place Observation instrument is designed to obtain information about the following:

1. The tasks which the worker in the workplace has to carry out.
2. The machinery and other equipment which he uses to assist him in carrying out these tasks.
3. The extent to which this machinery is manually or automatically controlled.
4. The requirements and demands of the job.
5. The job context (working conditions and environmental demands: covering such factors as temperature, vibration, noise, atmosphere, physical hazards, work schedule, relationships with other people).
6. Linkage to other workplaces.

METHODOLOGY

Information should be obtained principally through observation of the workplace by an investigator who should be familiar with the technological aspects of the industry. His observations can be supplemented by discussions with workers, with the departmental supervisor, with expert staff personnel in the plant, and with union specialists and representatives of other worker organisations. Discussions will act as a check that observations are correct and will help the investigator to gain a thorough understanding of the way work in the department is organised.

Criteria for identifying workplaces to be observed

If a number of workers are on identical jobs, on the same shift, with the same supervisor, then only one observation needs to be made and the data from this will be integrated with the attitudinal data from these workers.

If there is any major difference between one job and another, either in the job itself or in the work environment (e.g. supervision, shift), a separate work observation needs to be made.

SCALES AND CODES

1. Applicability code A.
   Does apply (1).
   Does not apply (0).

2. Scale of amount of time T.
   (a) Not present.
   (b) Rarely (this means approximately less than one-tenth of working time).
   (c) Occasionally (this means approximately one-tenth to one-third of working time).
   (d) Frequently (this means approximately a half to two-thirds of working time).
   (e) Continually (this means approximately two-thirds and over of working time).

3. Certain items have special scales and codes which are indicated in the text.

   The scales are only for estimation purposes. No time measurement should be done. The numbers are only given to operationalise what is meant if someone wants to use "rarely", "occasionally", "frequently", and so on. Imagine that these are adjectives without a clear meaning and that they are used only as an approximation. The variation scale does not need measurement either, only an estimation.

1. GENERAL DESCRIPTION OF WORK PROCESS (REPORT I)

(a) Describe briefly the work process and the machinery used for the unit being investigated. Plant documentation should be used to assist this where possible.

(b) Prepare a chart of the work area, designating every work place and, where applicable, setting out the sequence of the steps followed in the work process (see examples in Appendix 1).

(c) Describe the product or products made by the unit.

(d) Take the chart and indicate for the unit under investigation the nature of the workplace dependency, if any (i.e. the extent to which the job can be done in isolation or requires interaction with other workers at other workplaces). Indicate by lines and arrows the relationship between workplaces which have an interdependency. Give each workplace a number 1, 2, 3, 4, etc.

(e) Identify on the layout and flow chart all the workplaces to be observed.

(f) Prepare a list of job titles and give the number of the corresponding workplace on the layout and flow chart.

(g) Draw an organisational chart of the unit being investigated.

(h) State if the unit has a permanent group of skilled maintenance men (e.g. electricians) located in it or draws its maintenance workers from a separate maintenance unit.

2. DISTRIBUTION OF TASKS (REPORT II)

Describe briefly the job performed by the worker at his workplace:

(a) List the major job tasks.
(b) Estimate the approximate percentage of time devoted to each task by the worker.

Example:

Number of workplace ....
Job title ....

Name of machine(s) at workplace ....

A. TASKS	B. Percentage of the job
1. Operate machine	40
2. Set up machine	30
3. Consult foreman regarding job order	10
4. Stack and arrange materials and equipment	15
5. Perform preventive maintenance	5
	100

3. ANALYSIS OF WORKPLACE (REPORT III)

In order to analyse each workplace which has been selected for observation, it must be rated on a number of dimensions. The following pages set out the various dimensions and associated ratings.

The actual ratings of each workplace selected for observation should be recorded on the separate summary sheet/recording form at Appendix 3. A separate recording form should be completed for each selected workplace.

Time scale T

1. Not present
2. Rarely
3. Occasionally
4. Frequently
5. Continually

Scale of variation V

1. Proportion of time spent of the daily work activities does not vary within a week.
2. Proportion of time spent of the daily work activities significantly varies once a week.
3. Proportion of time spent of the daily work activities significantly varies more than once a week.

1. MENTAL REQUIREMENTS

1.1. Sources of Job Information

Rate each of the following sources of information indicating:

(a) what proportion of the day's activities of the worker requires the use of each source; and
(b) whether this proportion varies within a week.

1i __ T Written materials (words, e.g. job instructions, etc.)

ii __ V

2i __ T Quantitative materials (numbers graphs) from which details have to be selected, e.g. tables of numbers, technical drawings, etc.)

ii __ V

3i __ T Optical or acoustical signals, e.g. dials, signal lights, hand signalling, clocks, whistles, horns, etc.)

ii __ V

4i __ T Observation of the work process (tools, equipment, machinery, and other mechanical devices which are sources of information when observed during use or operation); (parts, materials, objects, etc., which are sources of information when being modified, worked on or otherwise processed or if not in process when being inspected, handled, packaged, distributed, or selected, etc. (such items or materials in inventory, storage or distribution channels, items being inspected, etc.).

ii __ V

5i __ T Verbal sources (consider only verbal communication which is relevant to job performance), e.g. verbal instructions, orders, requests, conversations, interviews, discussions, formal meetings, etc.).

ii___V

1.2. Job structure

6____S Job structure (Indicate, using the code below, the amount of "structure" of the job, that is the degree to which the job activities are "predetermined" for the worker by the nature of work, the procedures, or other job characteristics; the more highly structured jobs permit less deviation from predetermined patterns, and little or no need for innovation, decisionmaking, or adaptation to changing situations.)

S Code Amount of job structure

1. Virtually no deviation from a predetermined job programme (e.g. traditional assembly-line work, etc.).
2. Only moderate deviation from predetermined work programme is possible (e.g. stock handling, issuing material according to lists, planned maintenance (according to a programme,) etc.)
3. Considerable change from a programme or work routine is possible; work activities change considerably from day to day or even from hour to hour, but within some reasonable and expected bounds (e.g. traditional repair of automobiles, working on a common type of machine tool such as lathe, etc.).
4. Relatively little programmed work (routine); the job is characterised by considerable opportunity for improving methods, devices, etc., and the necessity for making decisions (e.g. supervision of production, fitting, and setting up the transfer line, etc.).
5. Virtually no established programme of activities; the position involves a wide variety of problems which must be dealt with; the solution to these problems allow for unlimited resourcefulness and initiative.

1.3. Information Processing Activities

This section refers to various human operations involving the "processing" of information or data. Rate each of the following in terms of its applicability to the completion of the job.

7____A Listing of information - recording (e.g. writing down of acoustic, optical or other information in records, work lists, report files, etc.).

8____A Classification, arrangement and combination of information to form other information (e.g. selection, compilation, and copying of special data from lists, records, and other written material; evaluation of charts of production schedules, inventory sheets; addition, subtraction of figures, calculation of machine input/output, rate of turnover of stock, etc.)

9___A Application and evaluation of information; selection of data from data collections, checking the possibility of application, and taking steps according to the selected data (e.g. use of literature (e.g. to determine cutting speed), determination of the production process on the basis of general outline of routing; processing according to work drawings; control of machine in accordance with description or diagram.

1.4. Use of Knowledge

10___S Level of general education (indicate the level of education generally or typically required of persons who are selected for this job; include education in elementary, high school, colleges, etc.; do not include technical or vocational school training - see item 11).

S Code Each country should formulate code(s) appropriate to its education system in order to arrive at a figure (not exceeding nine categories) comparing the average amount of education required of workers in the units being investigated.

11___S Level of vocational, technical education (indicate the level of vocational (technical) education generally or typically required of persons who are selected for this job); include technical middle schools, technical high schools, technical colleges, etc., do not include job training described in item 12.

S Code See note re item 10.
If questions 10 and 11 cannot be answered separately, information about both general and technical education might be given together.

12___S Job-related training, except for education described in items 10 and 11, i.e. the time necessary to become acquainted with the work, apprentice, mostly on the job and orientation training, etc.

S Code

0 None.
1. Up to 1 month.
2. More than 1 month to 3 months.
3. More than 3 months to 6 months.
4. More than 6 months to 12 months.
5. More than 12 months to 3 years.
6. More than 3 years.

2. PHYSICAL REQUIREMENTS

2.1. Use of Devices and Equipment

Consider in this category those devices which are used to move or modify or measure work pieces, materials, products or objects.

13___A Manually driven (e.g. hammer, saw, hand-measuring instruments, etc.).

14___A Power driven (e.g. manually controlled or directed devices using an energy source such as electricity, compressed air, fuel, hydraulic fluid, etc., in which the component part which accomplishes the modification is hand held, such as electrical hand drills, welding equipment, pneumatic screw drivers, etc., as well as devices small enough to be entirely hand held).

15___A Stationary machines/equipment used to process, fabricate, or otherwise modify parts, objects, materials, etc.

16___A Control devices on which the operator has to do only simple settings: hand- or foot- operated devices with distinct positions or definite settings.

17___A Control devices on which the operator makes variable setting of controls: hand- or foot-operated devices that can be set at the beginning of the operation, or, infrequently, at any position along a scale ("continual" settings).

18___A Keyboard devices, control panels, computer terminals, etc.

19___A Hand-operated transport and hoisting equipment (e.g. hand trucks, hand-driven cranes, etc.).

20___A Power driven transport and hoisting equipment (e.g. fork lifts, cranes).

21___A Remote controlled equipment (e.g. conveyor systems, etc.).

2.2. Body Positions Postures

Indicate by code the approximate proportion of working time the worker is engaged in the following activities.

22___T Sitting

23___T Standing (do not include walking).

24___T Walking (running) moving between machines.

25___T Climbing (e.g. stairs, ladders, on machinery).

26___T Kneeling/stooping: also crawling, crouching, or other related body positions which may be uncomfortable or awkward.

2.3. Level of Physical Exertion

Lifting.

27___T Over 40 kg. If item is coded 3 or above then go to questions 32-36.

28___T Over 25-40 kg. If item is coded 3 or above then go to questions 32-36.

29___T Over 15-25 kg. If item is coded 3 or above then go to questions 32-36.

30___T Over 5-15 kg. If item is coded 3 or above then go to questions 32-36.

31___T Under 5 kg.

Carrying.

32___T Over 25 kg. If item is coded 3 or above then go to question 37.

33___T Over 10-25 kg. If item is coded 3 or above then go to question 37.

34___T Over 5-10 kg. If item is coded 3 or above then go to question 37.

35___T Over 3-5 kg. If item is coded 3 or above then go to question 37.

36___T Under 3 kg.

37___T Pulling.

38___T Pushing.

39___T Reaching.

2.4. Manipulation Co-ordination Activities

Indicate by applicability code A whether the worker is engaged in the following activities:

40___A Finger manipulation - making careful finger movements in various types of activities: usually the hand and arm are not involved to any great extent (e.g. fine assembly work, use of precision tools, operating keyboard devices, etc.).

41___A Hand-arm manipulation: the manual control or manipulation of objects through hand and/or arm movements which may or may not require continuous visual control (e.g. placing components in position, repairing automobiles, etc.).

42___A Hand-arm steadiness: maintaining a uniform controlled hand - arm posture or movement (e.g. using a welding torch etc.)

43___A Eye-hand and/or eye-foot co-ordination: the co-ordination of hand/or foot movements where the movement must be co-ordinated with what is seen (e.g. driving a vehicle).

44___A Hand-ear coordination: the co-ordination of hand movements with sounds or instructions that are heard (e.g. listening for noise indicating fault in machinery and making adjustment).

45___A Visual stress (e.g. working with a computer terminal screen or other form of visual display screen: looking for small errors in moving components).

2.5. Other Work Requirements

This section lists various types of demands that the job situation may impose upon the worker, usually requiring that he adapt to these in order to perform his work satisfactorily. Rate the following items in terms of the amount of time they require or the number of times they occur.

46___T Specified work pace set by machine, (e.g. on assembly line, etc.)

47___T Repetitive activities: performance of the same physical or mental activities repeatedly without interruption for periods of time but without specified work pace.

48___S Lenght of work cycle.

S Code Length or work cycle

1. Under 1 minute.
2. 1-5 minutes.
3. 5-10 minutes.
4. 10-60 minutes.
5. More than 60 minutes.

49___T Being tied to a workplace: because of the nature and organisation of the work process, the worker cannot leave his place while working unless somebody else replaces him while he is away.

50___T Need to follow specific set procedures or routines (e.g. following check-out list to inspect equipment or vehicles).

51___A Time pressure of situation (e.g. urgent time deadlines, rush jobs).

52___A Probability of a wrong work decision having serious economic consequences for the plant (e.g. damage which cannot be repaired quickly, major machine breakdowns, plant stoppages).

53___T Attention to detail (e.g. checking tolerances, spotting defects).

54___T Vigilance: need to be aware of variations in a continually or frequently changing situation (e.g. watching frequently changing dials and gauges, etc.).

55___T Working under distraction (e.g. telephone calls, interruptions, disturbances from others, etc.).

56___A Up-dating job knowledge: need to keep knowledge current, being informed of new developments related to the job.

3. JOB CONTENT

3.1. Physical Working Conditions

Rate the approximate amount of time the worker is exposed to each condition.

57___T Air contamination (e.g. dust, fumes, smoke, oil vapour, toxic conditions, disagreeable odours, etc.).

58___T Vibration: vibration of whole body or body limbs as a result of operating machinery (e.g. operating an air hammer, compressed air chisel, etc.).

59___T Inadequate illumination: too dark, glare, etc.

60___T Dirty environment: an environment in which the worker and/or his clothing easily becomes dirty, greasy, dusty, etc.

61___T Wetness and/or humidity: working under conditions of humid air, water, oil on the floor, etc.

62___T High temperature which results directly from the work process: conditions in which the worker might experience severe discomfort or heat stress, as in boiler rooms, around furnaces, etc. Typically this would occur in a dry atmosphere at about 95° F (33° C) and in a humid atmosphere at about 80° F (27° C) or 85° F (30° C).

63___T Air conditioning.

64___T Awkward or confining work space or any condition which causes the worker physical discomfort.

65___S Noise intensity (indicate, using the code below, the typical noise level to which the worker is exposed).

S Code

1. Permits normal level of verbal communication.
2. Necessitates the raising of voice in order to be heard.
3. Some difficulty in hearing; raising of voice without gestures having to be used.
4. Uncomfortable noise level; must shout and gesture to communicate.
5. Painful noise level; communication almost impossible.

3.2. Physical Hazards

66___S Likelihood of accidents requiring medical attention.

S Code

1. Highly likely.
2. Not very likely.
3. Exceptional.

3.3. Relationships with Other Persons

67___S Contact permitted by technical work environment (indicate what possibilities the worker has for interpersonal communication in his work).

S Code

1. Other persons are available for communication within talking distance.
2. Other persons are available for communication within shouting distance only.
3. Other persons are available for communication only within signalling distance (visual and/or auditory signals which do not need technical means).
4. Other persons are not available for communication within human visual and/or auditory limits, but can be reached by technical media of communication (telephone, loudspeakers, teleprinters, etc.).
5. Other persons are not available for communication by any means; the workplace is completely isolated.

Frequency of job-required personal contact with:

68___T Co-workers.

69___T Supervisors.

70___T Others.

71___S Supervision normally required.

S Code

1. None
2. Moderate
3. Close

4. LEVEL OF AUTOMATION

4.1. Manual and Automated Activities at the Workplace (use special code below)

		Done by machine without human assistance	Done by machine with human assistance	Done manually	Does not apply (or is very incidental)
72	Feed of workpiece	___(3)	___(2)	___(1)	___(0)
73	Positioning of workpiece	___(3)	___(2)	___(1)	___(0)
74	Fastening of workpiece	___(3)	___(2)	___(1)	___(0)
75	Selection of tool	___(3)	___(2)	___(1)	___(0)
76	Setting of tool	___(3)	___(2)	___(1)	___(0)
77	Moving tool into working position	___(3)	___(2)	___(1)	___(0)

78	Start of machining conversion of material or workpiece (path information)	___(3)	___(2)	___(1)	___(0)
79	Progress of conversion of material or workpiece (path information)	___(3)	___(2)	___(1)	___(0)
80	Change of position of tool during work (switching information)	___(3)	___(2)	___(1)	___(0)
81	Change of wornout tools or change of tools for other reasons	___(3)	___(2)	___(1)	___(0)
82	End of conversion of material or workpiece	___(3)	___(2)	___(1)	___(0)
83	Release of workpiece	___(3)	___(2)	___(1)	___(0)
84	Removal of workpiece	___(3)	___(2)	___(1)	___(0)
85	Check of results of conversion of material or workpiece	___(3)	___(2)	___(1)	___(0)

4.2. Supplementary Measures

A supplementary proposal to measure the level of automation of the work tasks is being prepared by Dr. Wieser, revising a set of items developed by the Technical Group at Uniejow/Jablonna based on the Taylor classification of levels of technological sophistication. This will be forwarded to the national teams as soon as possible by the Vienna Centre and may be added to the WPO by the various teams if acceptable.

OPTIONAL

4.2. Level of Automation: Supplementary Information

86___S The extent to which the operator and machine are dependent on earlier task completion.

S Code : Extent of dependence on earlier task completion

1. The operation is a continuous process. Therefore the completion of the task at this workplace depends wholly on task completion at an earlier stage.
2. The completion of the task at this workplace can be carried out more or less independently once the task at an earlier stage is completed.
0. Does not apply.

87___S The extent to which the machine is dependent upon the operator.

S Code: Machine dependence on operator

1. The operator supervises a completely automated machine which can perform a production operation itself for sustained periods of time, and will shut down automatically in the event of trouble.
2. The operator supervises a machine which allows him to leave the controls for a limited period of time.
3. The operator supervises a machine he cannot leave at all during any portion of the task without stopping the work.
4. The operator works the machine himself and therefore cannot leave the machine without stopping the work.
0. Does not apply.

88___S Type of information received by the operator from the machine.

S Code : Type of information from machine

1. The machine itself controls the progress of the operation. Only malfunctions or the accomplishments of certain steps are indicated, e.g. through control lights.
2. The machine measures the progress of the operation. The operator receives the information from control panels and/or visual displays.
3. The operator himself has to control the work process using his experience and/or measurement devices (e.g. micrometer).
0. Does not apply.

89___S The proportion of routine operations which are automated:

S Code : Proportion of routine operations automated

1. Virtually all the routine operations at this workplace are automated.
2. The tasks at this workplace have a moderate proportion of routine operations which are not automated.

3. The tasks at this workplace contain a great proportion of routine operations which are not automated.

0. Does not apply.

Date ....Project ....Your initials ....

Person or Institution ....

Town ....

Subject ....

....

....

Summary of the discussion ....

....

....

....

....

Signature of the director

Date ....Project ....Your initials ....

Person or Institution ....

Town ....Tel. No............Minutes ....

Subject ....

....

....

Summary of the discussion ....

....

....

....

....

....

Signature of the director

Research Instrument V

WORKERS QUESTIONNAIRE

Opening statement to be prepared at discretion of each national team.

First, we would like to learn about the nature of your work.

1. What is the title of your job?

Now, would you rate various aspects of your job. For this purpose, we ask you to use a scale that looks like this:

To a little extent	1	2	3	4	5	6	7	8	9	To a large extent

For each of the items listed below check an appropriate place on this scale. For instance, if your job contains very little variety you would rate it somewhere on the left side - if it contains much variety you would rate it toward the right side. Please put your check mark under the appropriate number on each scale.

2. FIRST, TO WHAT EXTENT DOES YOUR PRESENT JOB CONTAIN:

2.1. Variety?
2.2. Independence?
2.3. Responsibility?
2.4. The chance to use your knowledge and training?
2.5. Possibility of working out new and better ways to do your job?
2.6. The need to learn new things?
2.7. Interesting work?
2.8. The chance to develop your own abilities?

3. HOW WOULD YOU DESCRIBE RELATIONS WITH YOUR CO-WORKERS?

TO WHAT EXTENT DO YOU AND YOUR COWORKERS:

3.1. Have a chance to talk with each other while working?
3.2. Help each other in getting work done?
3.3. Take a personal interest in each other?

4. HOW WOULD YOU DESCRIBE YOUR IMMEDIATE SUPERVISOR? TO WHAT EXTENT DOES HE:

4.1. Give sufficient explanations and instructions
4.2. Leave you alone unless you want help?
4.3. Listen to you when you have something important to say?
4.4. See you get justly rewarded for your work?
4.5. Have sufficient skills?

5. NOW WE WOULD LIKE TO KNOW ABOUT YOUR PAY AND OPPORTUNITIES IN THIS JOB

TO WHAT EXTENT DO YOU HAVE:

5.1. Good pay?
5.2. Job security?
5.3. Opportunity for upgrading or promotion?
5.4. Chance to improve your occupational skills?
5.5. Opportunity to get more general education?

6. FINALLY, WHAT ABOUT YOUR WORK SITUATION AND THE PRESSURES OF YOUR JOB?

TO WHAT EXTENT DO YOU HAVE:

6.1. An overload of work?
6.2. Physically strenuous work?
6.3. Mentally strenuous work?
6.4. A likelihood of job-related accidents and illness?
6.5. Convenient hours of work (starting and ending of your work day)?
6.5. Good physical conditions (such as light, temperature, cleanliness, low noise level, etc.)?

Now, thinking about the very things you have just described on your present job, we would like to know how <u>important</u> each of them is to you.

7. FIRST ABOUT THE NATURE OF YOUR JOB.

HOW IMPORTANT TO YOU IS THAT THE JOB CONTAINS:

7.1. Variety?

etc. (same items as under question 2).

8. Next, about the people you work with.

<u>HOW IMPORTANT TO YOU IS IT THAT YOU AND YOUR CO-WORKERS</u>:

8.1. Have a chance to talk with each other while working?
etc. (same items as under question 3)

9. HOW IMPORTANT TO YOU IS IT THAT YOUR SUPERVISOR:

9.1. Gives sufficient explanations and instructions? etc. (same items as under question 4).

Now, about your pay and opportunities.

10. HOW IMPORTANT TO YOU IS IT THAT YOU HAVE:

10.1. Good pay?

etc. (same items as under question 5).

11. GOING BACK TO THE QUESTIONS OF WORKING CONDITIONS AND PRESSURES OF YOUR JOB.

HOW IMPORTANT TO YOU IS IT THAT YOU DO NOT HAVE:

11.1. An overload of work

etc. (same questions as under question 6)

12. After this long list of ratings, we would like to know how satisfied you are, all in all, with your job (your present work at your present workplace). (Ans.: Very satisfied; Somewhat satisfied; Partly satisfied, partly not; Somewhat dissatisfied; Very dissatisfied.)

13. Would you like to have a different job than the one you have now (your present work at your present workplace)? (Ans.: Yes; No.)

14. Do you now see a real possibility of finding another job? (Ans.: Yes; Maybe; No.)

15. Do you actually intend to change your job during the next 12 months? (Ans.: Yes; Maybe; No.)

Thank you for this information about your job. Now we would like to know something about the management of this company.

16. On this sheet, will you check the statement which best reflects your over-all point of view on each of the five questions stated.

16.1. Are the equipment and resources with which the workers do their work at your plant efficient and well maintained?
16.2. Does the company try to improve working conditions?
16.3. Is the work process sensibly organised at this company?
16.4. Is the management of this company willing to consider the workers' ideas and suggestions?
16.5. Does the company take an interest in the personal welfare of the workers (e.g. pay, housing, individual problems, etc.)? (Ans.: To a very great extent; To a great extent; To some extent; To a little extent; To a very little extent.)

17. We are especially interested in what gives a person the best chance to improve his position (get promotion or get a higher rating) in your department.

Would you say that, for instance, skill has a lot to do with improving a person's position (rating) in your department, something to do with it, or nothing at all (Ans.: A lot to do with it; Something to do with it; Nothing at all; I don't know.)

17.1. Skill (working ability)?
17.2. Working hard?
17.3. Seniority in the company?
17.4. Being a favourite of the boss or supervisor ("pull")?
17.5. Union influence?
17.6. "Social activity?"

18. Can you please tell us something about relations between the people that work in your plant? (Ans.: Regularly; Occasionally; Never.)

Do you have informal talks with:

18.1. Co-workers?
18.2. Any members of the supervisory and administrative staff?
18.3. Do you and the supervisors of your unit have your meals or snacks together at the plant?
18.4. Does the company organise joint leisure time activities (trips, sporting activities, etc.) for you and for the members of the supervisory and administrative staff?
18.5. Do any of your co-workers invite you to their homes?
18.6. Do any of your supervisors or the administrative staff invite you to their homes?
18.7. Do you think that the members of the supervisory and administrative staff look down on you, because you are a worker?

19. Do the workers of your plant consider it possible to criticise the public and in the presence of persons concerned: (Ans.: Yes; No.)

19.1. Somebody of the rank-and-file workers?
19.2. Their foreman?
19.3. Shop engineers?
19.4. Head of your department?
19.5. Plant managers?

I have already explained that this study is about automation and introduction of new machinery. Let us go to questions directly concerned with this.

20. Is it the usual practice in the plant for workers to be informed prior to the introduction of new, more advanced machines and equipment: (Ans.: Yes; No.)

20.1. By management?
20.2. By union or other worker representatives?

21. 21.1. Has new machinery been introduced into your department while you have been working there? (Ans.: Yes; No.)
    21.2. If yes, were you officially informed about it beforehand? (Ans.: Yes; No.)
    21.3. If yes, how long before the installation was the information given? (Ans.: Less than 1 month; 1-3 months; 3-6 months; More than 6 months.)

22. 22.1. Have you made any suggestions to management for changing the machinery or work process during the past year? (Ans.: Yes; No.)
    22.2. If yes, was any suggestion accepted? (Ans.: One or more suggestion was accepted; No suggestion was accepted; The suggestion is still being considered; I don' t know.)

23. Do you think that the workers in your plant have directly or indirectly influenced the decisions concerning the introduction of new machinery? (Ans.: Yes; No; I don't know.)

24. In your opinion, how does automation generally influence: (Ans.: Increase; No change; Decrease; I don't know.)

    24.1. Job security?
    24.2. Participation of workers in decisionmaking?
    24.3. Prospects for young workers?
    24.4. Physical strain?
    24.5. Mental and nervous strain?
    24.6. Job satisfaction?
    24.7. Importance of unions for workers?

25. How has the introduction of new machinery affected the workers in this plant? (Ans.: For better; For worse; Made no difference; I don't know.)

26. 26.1. If work-saving, new machines and equipment are introduced into your plant would you:

    (a) Actively support such changes?
    (b) Quietly approve such changes?
    (c) Quietly disapprove of such changes?
    (d) Actively resist such changes?

    26.2. Why would you do so?

27. Are you afraid you might lose your own job because of automation? (Ans.: Yes; No.)

28. If your present job is taken over by new machines do you think you will be provided with other acceptable work? (Ans.: Yes; Possibly; No.)

29. Now we would like to know your opinion about the participation of workers in decisionmaking at the plant. First, which of the following statements best reflects your opinion about the opportunity workers should have? (Check one.)

    Workers should have the opportunity to participate in making decisions on all matters affecting the plant.

Workers should have the opportunity to participate in making decisions only on problems affecting their work.

Workers should have the opportunity to participate in decisions only when management asks them.

Workers need not have the opportunity to participate in decisionmaking.

30. Which of the following statements reflects your own willingness to participate in making decisions in your plant? (Check one.)

I am willing to take part in most decisions affecting the plant.
I am willing to take part in decisions affecting my work.
I am willing to take part in decisions only when management asks me.
I am not willing to take part in decisionmaking.

31. Is it possible for the workers of the plant to participate directly, or through their representatives, in decisions on the following questions: (Ans.: Yes; No; Don't know.)

31.1. Development of production plan?
31.2. Introduction of new machinery and equipment?
31.3. Work organisation and working conditions?
31.4. Personnel planning (planning of future work force demands)?
31.5. Determination of basis and methods of wage payment?
31.6. Workers' job classification and pay scale?
31.7. Determination of bonuses and other incentives?
31.8. Hiring of workers?
31.9. Promotion of workers?
31.10. Selection of workers for training courses?
31.11. Transfer to other work places?
31.12. Allocation of overtime work?
31.13. Disciplinary measures?
31.14. Lay-off and discharge of workers?
31.15. Cut-back of production and closing down of plant?
31.16. - 31.20. Optional items according to national interest?

32. Here are some ways that workers in different countries use to influence managerial decisions. Please check any activities in which you personally participated in the last two years. (Ans.: Yes; No; Don't know.)

32.1. Participation in the meetings of worker organisations (e.g. worker meetings, trade union meetings, meetings of the plant political party organisation, youth organisation).
32.2. Participation in the work of different committees at the plant (e.g. trade union, workers council, self-management committees, joint union-management committee).
32.3. Expressing your ideas to your immediate supervisor or management.
32.4. Expressing your ideas to your workers' representatives in the plant.
32.5. Contacting worker leaders outside the plant (e.g. higher level of trade union, self-management organisations).

32.6. Contacting the state (governmental) bodies.
32.7. Arousing public attention (e.g. organising demonstrations, distribution of leaflets, newspapers).
32.8. Methods of collective pressure (e.g. refusal of overtime work, slow-down, work to rule, strikes).
32.9 - 32.10. Optional items according to national interests.

Now we would like to ask you something about union activities.

33. In your opinion how important should each of the following issues be in union activities: (Ans.: Very important; Somewhat important; Not important.)

33.1. Improving wages, pensions, etc., of the workers?
33.2. Developing cultural and educational activities in the plant?
33.3. Providing recreational facilities (including vacations)?
33.4. Improving productivity?
33.5. Awakening a strong political attitude and consciousness among the workers?
33.6. Improving the terms and conditions of work?
33.7. Influencing the general social and economic policy of the country?
33.8. Handling worker grievances (complaints) and disputes?
33.9. Ensuring job security for workers?
33.10. Changing work organisation to increase job satisfaction (humanisation of work)?

34. How much attention does your union pay to the following issues: (Ans.: Too little; Adequate: Too much.)

31.1. Improving the wages, pensions, etc., of the workers?
etc. (same item as question 33).

35. All in all, how satisfied are you with your union? (Ans.: Fully satisfied; Somewhat satisfied; Partly satisfied, partly not; Somewhat dissatisfied; Very dissatisfied.)

We have asked you a lot of questions about your work, the plant, and union activities. Now we would like to know what you think about some public issues.

36. How concerned are you personally about the following public issues: (Ans.: I am not concerned; I am somewhat concerned; I am much concerned; I am very deeply concerned.)

36.1. Problems of preventing war and achieving peace?
36.2. Development of science and technology?
36.3. Problems of the country's economy?
36.4. Problems of the educational system?
36.5. Problems of cultural life?
36.6. Housing problems?
36.7. Crime and delinquency?
36.8. Health care?
36.9. Pollution?
36.10. Youth problem.

37. Do you think you could contribute effectively to a solution of the above issues? (Ans.: Yes, to most of them; Yes, to some of them; No to none of them; Don't know.)

38. If yes, by what means (Check the following):

38.1. Talking to your trade union representatives?
38.2. Speaking at union meetings?
38.3. Participating in demonstrations?
38.4. Working actively in committees?
38.5. Talking to party officials?
38.6. Talking to officials of the public administration (government officials)?
38.7. Sending letters to deputies (legislators) at national and local levels?
38.8. Participating in strikes?
38.9. Sending letters to newspapers?
38.10. Securing resolutions from organisations of which you are a member?
38.11. Other means (specify).

39. 39.1. Have you yourself done any of these things? (Ans.: Frequently; A few times; Once; Never.)
39.2. If "frequently" or "a few times", which of these things did you do? (To interviewer: using numbers from questions 38.1. - 38.11. for reference, check those things done frequently or a few times.)

40. Do you regularly take an active part in any of the following organisations (without pay) (Ans.: Yes, No.)

40.1. Political party?
40.2. Trade union?
40.3. Youth organisation?
40.4. Municipal organisations?
40.5. Cultural organisations?
40.6. Social health organisations?
40.7. Sports organisations?
40.8. Education organisations?
40.9. Institutions maintaining public order in the community?
40.10. Religious (church) organisations?
40.11. Other institutions?

41. 41.1. Only for the countries with planned socialist economy.) Do you participate in socialist emulation? (Ans.: Yes; No.)
41.2. If yes, what main obligations have you assumed for the current year?

FINALLY, JUST A FEW BITS OF INFORMATION ABOUT YOUR PERSONAL BACKGROUND

42. What is your age? (Ans.: 20 or under; 21-25; 25-30; 31-40; 41-50; 51-60; Over 60.)

43. Sex.

44. Education (general): according to national specifics.

45. Education (technical): according to national specifics. Job training not to be included.

46. 46.1. Do you study at present? (Ans.: Yes; No.)
46.2. If yes, where do you study? (Optional)

47. 47.1. Did you receive training for your present job? (Ans.: Yes; No.)
47.2. If yes, how long was it?
(Ans.: Up to 1 month; more than 1 month to 3 months; More than 3 months to 6 months; More than 6 months to 12 months; More than 12 months to 3 years; More than 3 years.)

48. 48.1. Did you receive training in connection with any of your previous jobs? (Ans.: Yes, No.)
48.2. If yes, on average how much training did you receive in your previous jobs? (Up to 1 month; More than 3 months to 6 months; More than 6 months to 12 months; More than 12 months to 3 years; More than 3 years.)

49. Do you receive any training now? (Ans.: Yes; No.)

50. How long have you been with this company?

51. How long have you been on your present job?

52. 52.1. Have you been unemployed during the last five years?
52.2. If yes, for how many months?
52.3. How much was in the past year?

53. What is your pay (without overtime) (either hourly, weekly, or monthly)?

54. 54.1. Are you a member of a trade union? (Ans.: Yes; No.)
54.2. If yes, which trade union?

2. Short Description of the Optional Study and Associated Studies

1. OPTIONAL STUDY: "AUTOMATION AND INDUSTRIAL WORKERS IN THE STEEL INDUSTRY"

1.1. The Optional Study "Automation and Industrial Workers in the Steel industry" is carried out by ten countries:

Austria:	Fred Margulies	Gewerkschaft der Privatangestellten Vienna
	Friedrich Prammer	Voest-Alpine AG Linz/Donau
	Georg Wieser	Institute of Sociology University of Vienna
Federal Republic of Germany:	Guenther Friedrichs	Industriegewerkschaft Metall, Frankfurt/Main
	Hans Mittler	ISO Institut Saarbruecken
	Guenter Sieber	Hoesch-Huettenwerke AG Dortmund
Finland:	Tuula Heiskanen	University of Tampere
	Liisa Rantalaiho	University of Tampere
France:	Yvette Lucas	University of Toulouse
Great Britain:	Chris Schumacher	British Steel Corporation London
Hungary:	Lajos Héthy	Institute of Sociology Hungarian Academy of Sciences Budapest
Japan:	Kenji Okuda	Nippon Kokan K.K. Tokyo

Sweden:	Ulf Aberg	Laboratory of Industrial Ergonomics, Stockholm
	Lena Martensson	Laboratory of Industrial Ergonomics, Stockholm
U.S.A:	S. Lee Spray	Dept. of Sociology Kent State University
Yugoslavia:	Valentin Jež	Self Management Research Centre, Ljubljana

1.2. The study deals with milling processes in steelworks, and comparing higher and lower levels of automation in slabbing mills and hot strip mills. Instead of a slabbing mill, some countries choose the continuous casting process for slab production, comparing it with conventional ingot casting. The process areas which were subject to research are shown in the diagram below.

In this basic approach the optional study had the same scope, used the same methods and the same instruments as the standard study (interviews with management and trade union representatives, work-place observations, and workers questionnaire). Modifications were kept to a minimum to ensure comparability with the standard study results. By agreement of participating countries the same major hypotheses were adopted. The different levels of automation were identified as "manual operation", "remote control", and "computer control", with comparison to be made between two or, wherever possible, all three of them.

The data collected in various steelworks of the participating countries were put onto punched cards, which were sent to the Toulouse University Computing Centre to be processed. Analysis methods were also kept close to those applied with the standard study. Good comparability between the standard and optional study results was aimed at throughout the study and may justly be expected to show when the reports are completed.

The study is co-ordinated by Mr. Fred Margulies (Austria), from whom further information can be obtained. His address is:

Gewerkschaft der Privatangestellten
Deutschmeisterplatz 2
A - 1013 Vienna

2. ASSOCIATED STUDY: COMPUTER SYSTEMS AND WORK DESIGN

2.1. Six countries participated in this study; several academic disciplines, as well as trade unions and management, were represented.
The principal researchers were:

Austria:	Fred Margulies	Gewerkschaft der Privatangestellten, Vienna
	Georg Wieser	Institut of Sociology University of Vienna
Denmark:	Flemming Agersnap	Institute of Organisation and Industrial Sociology Copenhagen
	Niels Bjørn Andersen	idem
	Henrik Bulow	idem
	Eva Schrøder	idem
	Erik Winther	idem
	B. F. Sørensen	
Federal Republic of Germany	I. Brandt	
	Ulrich Briefs	Wirtschafts- und Sozial wissenschaftliches Institut des Gewerkschaftsbundes Duesseldorf
	Klaus Hermann	Institut fuer Sozialforschung Duesseldorf
	M. Papadimitrou	idem
France:	William Grossin	Université de Nancy II
	Andrès Sole	idem
Great Britain:	Pauline Harris	Manchester Business School University of Manchester
	Dorothy Mercer	idem
	Enid Mumford	idem
Sweden:	Bo Hedberg	
	Olaf Edström	University of Gothenburg
	Gunhold Lidström	

2.2. The Computer Systems and Work Design Study was concerned with a different technology from that of the standard study, namely computer based information systems. The performance of these systems has not always come up to expectations, and some suggest that their potential has not been fully utilised. Social scientists and others believe that this is partly because too little attention has been paid to the role of man in the design and operation of computer applications, and there has been too ready an acceptance of technologically determined consequences for organisations.

The objective of the study was to extend our understanding of the process of designing and introducing computer based systems, and of the impact of such systems on the people involved with them, by seeking to relate aspects of the system design process and of the technology itself to features of the work structure and patterns of employee attitudes in a number of user situations. At the same time there was a more practical aim, to identify some alternatives which exist for system designers, and to show how computer based systems can be used to raise, rather than lower, the quality of working life for the user.

The study was carried out in two areas of industry-the banking sector and steel manufacturing; in each of the countries represented, computer based systems in one or both of these industries have been investigated. In order to observe the consequences for significant numbers of non-managerial users, particular types of systems were selected; in the case of banking, systems for data entry and the daily processing of customer accounts; and in the case of the steel industry, systems for the entry and processing of orders and production planning.

The main groups of users dealt with were, therefore, clerical workers in branch banking, and in order and production planning departments of steelmaking organisations. Within these boundaries, the researchers studied the most technologically advanced systems available in each country. In all cases these were established systems and the study was retrospective in so far as it dealt with their origination and development.

The study group followed the researchers in the standard study in considering that the social and economic structure in each country was a background variable which must be taken into account, affecting as it did both events at the organisational level and the behaviour and attitudes of individuals. The system design process, the principal independent variable, was analysed in terms of the objectives set, the participants in design and their values, and the procedures by which solutions were reached. The technology itself was described and classified. Data was also collected on a series of structural and ideological features of the organisations studied, and on the social characteristics of the user groups.

The major part of the information was collected by face-to-face interviewing using instruments jointly designed by the group. These included detailed questionnaires for members of senior management, systems designers, trade union officials, and departmental managers, dealing with the history of each system studied, and eliciting a current evaluation of the systems according to criteria appropriate to these groups of interviewees. Observation was used within the user departments to obtain systematic descriptions of the work structures. And the non-managerial users were interviewed about their views of the implementation of the system, its impact on their work situation, and their current level of satisfaction both with their job and with the system.

This study is co-ordinated by Mrs. Enid Mumford (Great Britain) from whom further information can be obtained. Her address is:

Manchester Business School
University of Manchester
Booth Street West
Manchester M15 6PB

3. ASSOCIATED STUDY: COMPUTER INFORMATION SYSTEMS AND MANAGEMENT

3.1. The following countries have carried out empirical investigations:

Austria:	Georg Wieser	Institute of Sociology University of Vienna
Denmark:	Niels Bjorn Andersen	Institute of Organisation and Industrial Sociology Copenhagen
	Poul H. Pedersen	idem
Federal Republic of Germany:	Herbert Kubicek	University of Cologne
Heiko Lippold		idem
	Elmar Reindl	idem
Great Britain:	Leela Damodaran	Loughborough University
	Ken D. Eason	idem
	Tom F. M. Stewart	idem

Poland (Adam Sarapata) and the United States (Daniel Robey) have expressed interest without having carried out empirical investigations.

3.2. The aim of this research project has been to increase insight into the constraints and the contingencies imposed by electronically based management information systems on the individual managers and the management structure. It is hoped that such an increased insight may make it possible for information specialists and managers to design systems and structures more conducive to efficiency as well as to job satisfaction.

The prime focus of the research has been to investigate the extent to which implemented computer based information systems have changed the tasks, the communication, the power, and the discretion of the individual managers.

Furthermore, important structural changes as a consequence of the introduction of computer information systems are studied especially focusing on changes in the relationship between line and staff functions.

In order to achieve this objective, substantial amounts of data have been collected on the tasks and changes in the tasks of the managers, their use of the system, their information needs, their evaluation of the Computer Information System (CIS), their evaluation of ease of use, their evaluation of use support, the impact of the CIS on two selected tasks of the manager, the distribution and change in distribution of influence in the organisation, level and change in methods and form of influence, level and change in power bases of individuals, level and change in discretion of individuals, attitudes towards CIS in general, user involvement in system design, external influence on the system design process, and the impact of CIS on responsibility, authority, and delegation.

The research has been carried out as case studies. In each country one or two companies (banks, manufacturing industries, hospitals, and mail-order companies) have been studied intensively. In each company at least two company functions and one integrated computer system have been studied.

Investigations have primarily been carried out with questionnaires, two interviews and a self-completion questionnaire per manager, subordinate questionnaire and a system designer questionnaire. Between ten and fourty managers have participated in each company.

First reports written on purely national data appeared in 1976, and the final integrated publication reporting on the main findings of the study is scheduled to appear in 1978.

The study is co-ordinated by Niels Bjorn Andersen (Denmark). Further information may be obtained from:

Information System Science Group
Copenhagen Business School
60 Howitzyej
DK-2000 Copenhagen F

3. List of Selected Documents and Papers Prepared During the Research*

1. Automation and Working Class. (Draft project prepared by A. Schaff (Vienna Centre) and A. Sarapata (Poland), 1969.

2. Motivation in Automated Society by the Hawaii University Research Team (USA) for the Conference in Moscow, 10-20 May 1971, January 1971.

3. Labour and Professinal Characteristics of Workers, Engineering and Technical Staff in Automated Production. (Project of the Institute of International Workers' Movement, Academy of Sciences of the USSR, April 1971.

4. Some Remarks on the Economic Problems of Automation by G. Torri, CISL (Italy), January 1972.

5. The Effects of Automation on Job Content. Review of relevant literature as contribution to the Moscow Conference of the international research project "Automation and Industrial Workers", by J. Rezler (USA), AUGUST 1972.

6. Final Research Design. Edited by the working group meeting at Neuwaldegg (Austria), January 1973.

7. On the Analysis of Importance Ratings by J. Forslin (Sweden), September 1974.

8. Speculation on Technological Comparability by S. C. Mills (Vienna Centre), September 1974.

9. Notes on Technology by L. Hethy (Hungary), November 1974.

10. A Statistical Design for International Comparisons by J. Forslin (Sweden), February 1975.

11. Sampling, Population, Inference, Comparisons by J. Rehak (CSSR), May 1975.

12. Sociological Properties; Their Representation, Measurement and Processing (Terminology and Conceptual Structure) by J. Rehak (CSSR), May 1975.

13. The Two Group Notion. A comment on US data by G. W. England (USA), May 1975.

14. Task and Movement Monotony. Some explorations with Finnish data, by L. Rantalaiho (Finland), May 1975.

15. Technology - A Three-dimensional Concept by L. Hethy (Hungary) and C. Mako (Hungary) for the Lauf Conference, May 1975.

16. The Ideology of Progress as an Artefact in Comparative Sociology (rough translation). Unrevised version by W. Grossin and A. Sole (France), April 1975.

17. Canonical Analysis of WPO Cards from Three Countries (FRG, Sweden, USA) by the Finnish Team, February 1976.

18. A Model of Generating and Testing Hypotheses. Chung si Ahn's (USA) comments as corrected by the University of Hawaii Group, February 1976.

19. To Multiply or to Subtract - this is the Question. Some separatist remarks in methodology concerning the following: how to jointly use responses to the worker questionnaire "To what extent" and "How important", by A. Sarapata (Poland), April 1976.

20. Proposals for Primary Information Analysis according to Research Instrument IV (Workplace observation) by the USSR Team, April 1976.

21. Suggestions about Unit Type, Job Type, and Automation Level Measures for Purposes of International Comparisons by G. W. England (USA), July 1976.

22. Some Remarks about Comparability of Data by M. Prussak (Poland), September 1976.

* Papers can be requested from the authors or the Vienna Centre and are available if the author agrees.